ETO Oral Questions and Answers

Steve Richards

April 12, 2025

ETO Oral Questions and Answers

Fifth edition

Dedicated to Jessica Brooks.

Contents

1 Introduction

This book tries to fill the need for Electro-Technical Officer (ETO) trainees about to sit for their CoC professional interview. If you have been a cadet officer studying for three years or are a time served electrical officer who wishes to obtain the ETO CoC, it can seem like an immense memory test. You need to remember a vast amount of knowledge and recall it with certainty and confidence in front of an expert examiner!

To help you consolidate your knowledge this book offers hints and tips gained from teaching cadets to succeed at this exam.

We present the MCA ETO syllabus then offer sample questions and answers that map across this syllabus. We also offer a series of mock interviews.

It is helpful to note that although the examiner can ask you ANYTHING in the syllabus, due to time constraints, the examiner will concentrate on what is deemed the most important areas. This means although you could be asked questions on first aid, you probably will not be asked about it. I do not need to cover this area in this book as an approved first aid certificate is required to sit for the CoC.

Note: Questions which require a long answer and a diagram, are placed facing each other to allow you to see the diagram and answer text simultaneously. This means that the page before will be blank. This is by design. Answers that will not fit in the space remaining on a page will start on the next page, subject to the above.

Please read page 160 to find out how you should be reading this book.

Next are the questions and answers.

2 Questions and answers covering the syllabus

Question 1.

Draw a ships electrical distribution system.

Answer:

- Draw the circuit of your last vessel.

  You draw the distribution diagram of your last vessel, if it was a HV ship then you draw the HV diagram. See the example Low Voltage diagram in figure 1 and a High Voltage diagram in figure 2

  If you have *NOT* sailed on a HV ship before, the examiner may still ask to draw a typical HV distribution system like those you learnt on your HV course. This is because when you pass the CoC you will be able to sail as ETO on an HV vessel.

  Ensure you have the correct number of generators, switch boards, that each generator has a breaker, each transformer has a breaker on each side and each major load has a breaker. A bus tie breaker must be shown. If your vessel was a DP ship or a passenger ship with HV and just a few HV loads, draw them in. Show the emergency generator(s) and shore side connection, if fitted. Add on voltages, watts generated and name the key breakers.

  The examiner will know what your last ship was from your discharge book, NoE application and your training record book if you completed one.

  Do not make things up or bluff, you are sitting in front of an expert surveyor who, daily, inspects ships and knows if you are bluffing!

  You do not have to draw a perfect and complete diagram, just enough to demonstrate your knowledge. That you understand the key aspects to the electrical supply and distribution of your last vessel. Practice drawing these diagrams dozens of times.

  Be prepared to answer questions about your drawing. Such as:

- Why do we have a bus tie breaker (so we can split or isolate the bus during faults or maintenance)?

- Why do we have 4 generators (multiple generators increase reliability and allow for economic variations in load), *the examiner will ask about your installation* and SOLAS informs us we need at least two.

- Why do we have an emergency generator and switchboard? (The emergency switch board provides power for essential auxiliaries used to get the main generators up and running plus essential loads including: navigational lights, radio communications, fire pump, fire alarm and steering gear. SOLAS informs us we need an emergency power source).

- What are the resistors fitted to each HV generator and why do LV generators not have them? (Please read pages 37 and 38)

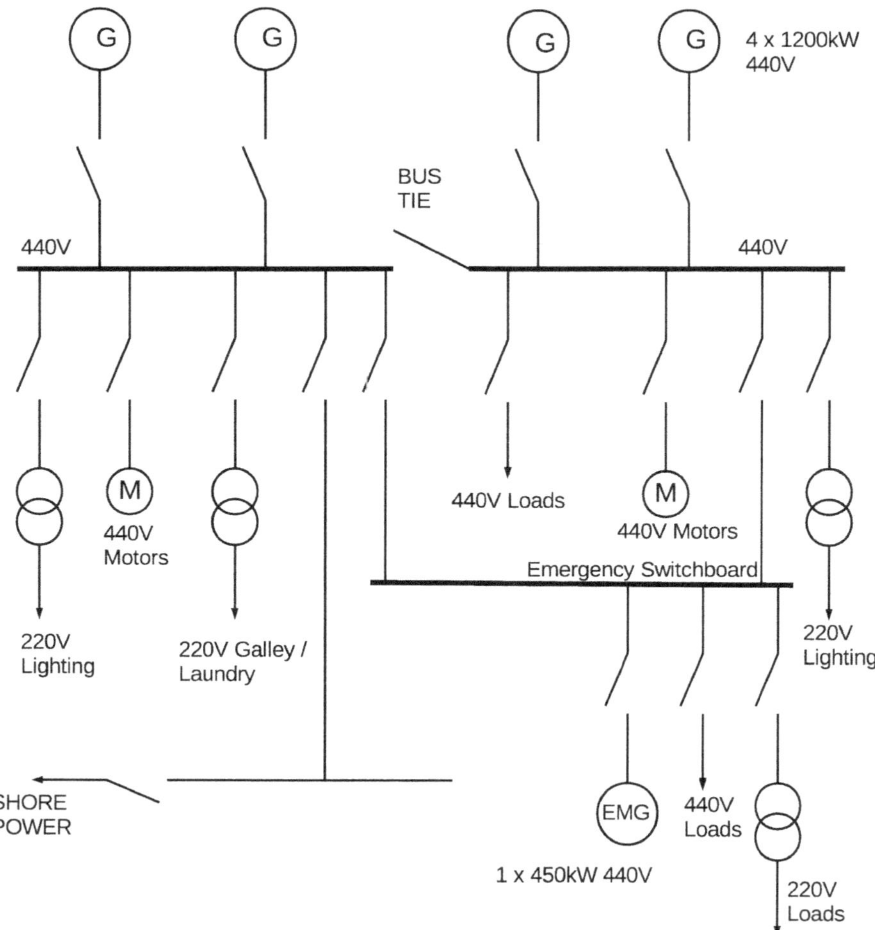

Figure 1: *Sample LV distribution diagram suitable to answer your first question if your ship had LV.*

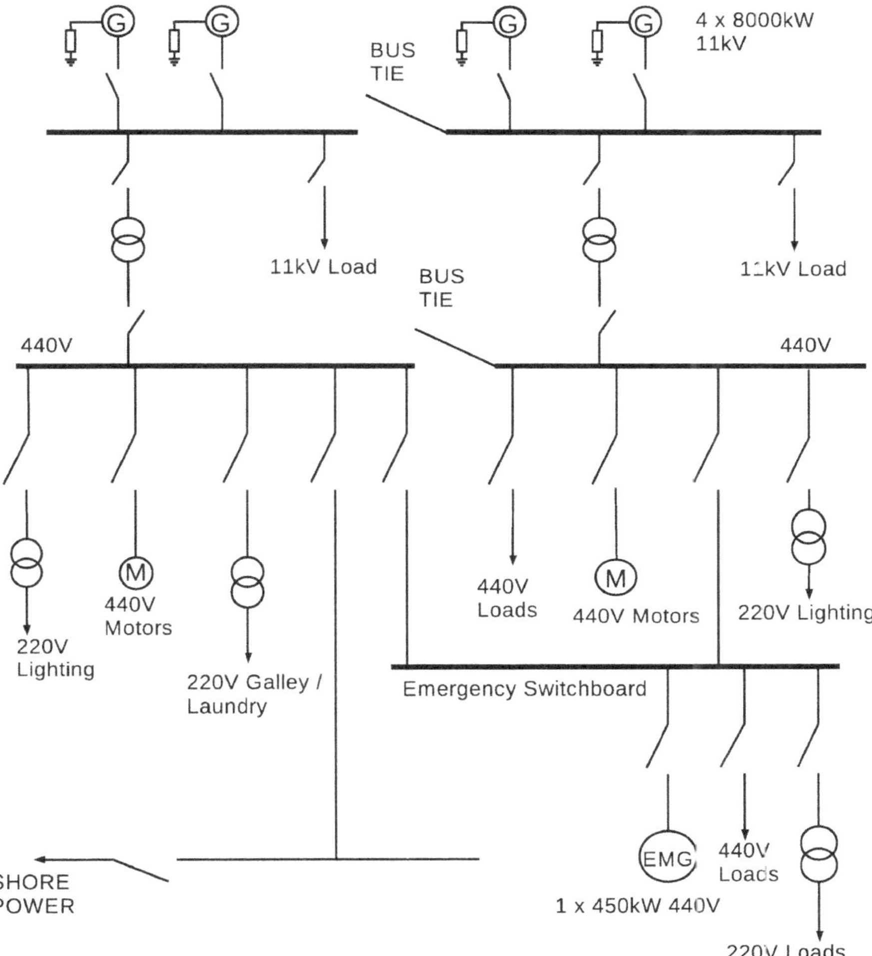

Figure 2: *Sample HV distribution diagram suitable to answer your first question if your ship had HV.*

Question 2.

What happens during a blackout?

Answer:

- Transitional batteries automatically supply: emergency lighting, radio communications, internal communications and alarm systems for a minimum of 30 minutes or until the emergency generator comes online.

Question 3.

How long does the emergency generator have to start?

Answer:

- Within 45 seconds it must be providing power to the emergency switchboard.

Question 4.

How long does it need to run for?

Answer:

- On a cargo vessel 18 hours,
- On a passenger vessel 36 hours.

Question 5.

On your last vessel what were the starting methods for the emergency generator?

Answer:

- The emergency generator had two independent methods: air start from its own air receiver and 12 Volt battery start. *You describe your installation.*

Question 6.

Where did the air come from?

Answer:

- The main air receivers. If these were empty, a hand started, diesel powered air compressor could provide air for the local receiver. *You describe your installation.*

Question 7.

What type of battery did your emergency generator use?

Answer:

- Lead acid, two banks of 12V batteries, each bank connected by a A/B changeover switch. *You describe your installation.*

Question 8.

What is a good way to verify that a lead acid battery is charged and in good condition?

Answer:

- I would use a battery tester that caused the battery to deliver some current through the tester and measure the battery voltage on load.

Question 9.

Why do you say that? Why not just measure the terminal voltage?

Answer:

- The terminal voltage of a lead acid battery does not tell you much about the battery condition. By forcing the battery to deliver some current, it indicates the value of its internal resistance. As the battery ages its internal resistance increases, reducing the onload voltage.

Question 10.

How often should the emergency generator be tested?

Answer:

- Once per week offload, once per month onload.

Question 11.

How long should the emergency generator be run for?

Answer:

- Typically 30 minutes, until the temperatures stabilise and indicate normal operation.

Question 12.

Okay, you have one main generator driving the board and you want to put another generator online, tell me how you would do it.

Answer:

- I would ask the EOOW if it was okay to start the chosen generator.

- With permission I would start it from the generator control panel on the main switchboard.

- I would wait for it to stabilise, observe pressures and temps, adjust the frequency to be between 60.5 to 62 Hz and the voltage to be within plus or minus 5% of the bus voltage.

- I would switch the synchroscope on, adjusting the incoming frequency to be slightly faster than the bus. This is indicated by the synchroscope turning slowly clockwise.

- When the synchroscope is at 11 o'clock, I would press the breaker close button. Immediately the incoming generator will start to deliver some power.

- If I wanted to share the load equally, I would increase the fuel/speed of the lightly loaded machine and reduce the fuel/speed of the highly loaded machine. This way I can keep the frequency constant.

Question 13.

Why did you press the close button at 11 o'clock?

Answer:

- There is a delay between pressing the button, the breaker starting to move and the breaker making contact.

- If I pressed the button at 12 o'clock, by the time I reacted and the breaker closed, the incoming generator would be out of sync with the bus and the possibility exists of damage being caused.

Question 14.

Tell me the name of the most common electric motor on board vessels.

Answer:

- Induction motor.

Question 15.

Can you describe how they work?

Answer:

- An induction motor consists of a rotor constructed out of solid bars mounted as a cylindrical cage, with the bars shorted together at each end.

- The rotor cage is mounted on a shaft that is free to rotate.

- The rotor cage fits into the cylindrical three phase stator or field windings.

- Three phase is fed to the stator windings creating a rotating magnetic field.

- This rotating field cuts the stationary rotor cage bars inducing an EMF.

- This EMF causes a current to flow along the rotor bars.

- This current creates a magnetic field around each bar which interacts with the rotating magnetic field.

- Therefore the rotating magnetic stator field forces the rotor to rotate.

Question 16.

Does an induction motor rotate at synchronous speed?

Answer:

- No, it must rotate slightly slower, to create the magnetic field cutting action.

- This difference in speed is called *slip*, typically *slip* is around 5%.

Question 17.

What is the starting current of an induction motor?

Answer:

- Typically between 4 and 7 times the normal full load current, depending upon the design.

Question 18.

How can we reduce this high starting current?

Answer:

- Commonly we can use Star-delta, auto-transformer and soft starters (variable speed drives).

Question 19.

Can you describe the Star-delta system?

Answer:

- You draw a sketch something similar to that shown in figure 3 and explain its operation, similar to the following:

- Three phase power enters on L1, L2 and L3.

- When the MCCB is closed, voltage is stepped down via the transformer T1 to energise the control circuit.

- Contactor K1 remains de-energised as the control voltage can not get through the open circuit start button.

- When START is pressed, the control voltage now flows through K1 coil and energises it.

- K1 closes, T1 actuates closing K2, the motor is now connected in star configuration and spins up.

- K1 auxiliary contact also closes across the start button allowing you to remove your finger from the start button.

- After a few seconds T1 timer times out, causing K2 to open and K3 to close.

- The motor now runs in delta configuration.

- K2 and K3 also have a physical interlock preventing them from closing simultaneously.

- Pressing the STOP button *open circuits* the control circuit causing K1 and K3 to de-energise, stopping the motor.

- An overload sensor OL will cause the OL switch to open preventing power flow through the control circuit, causing all contactors to open, stopping the motor, if it gets overloaded.

- See figure 4 for a view of the phase voltages. In star it is about 60% less.

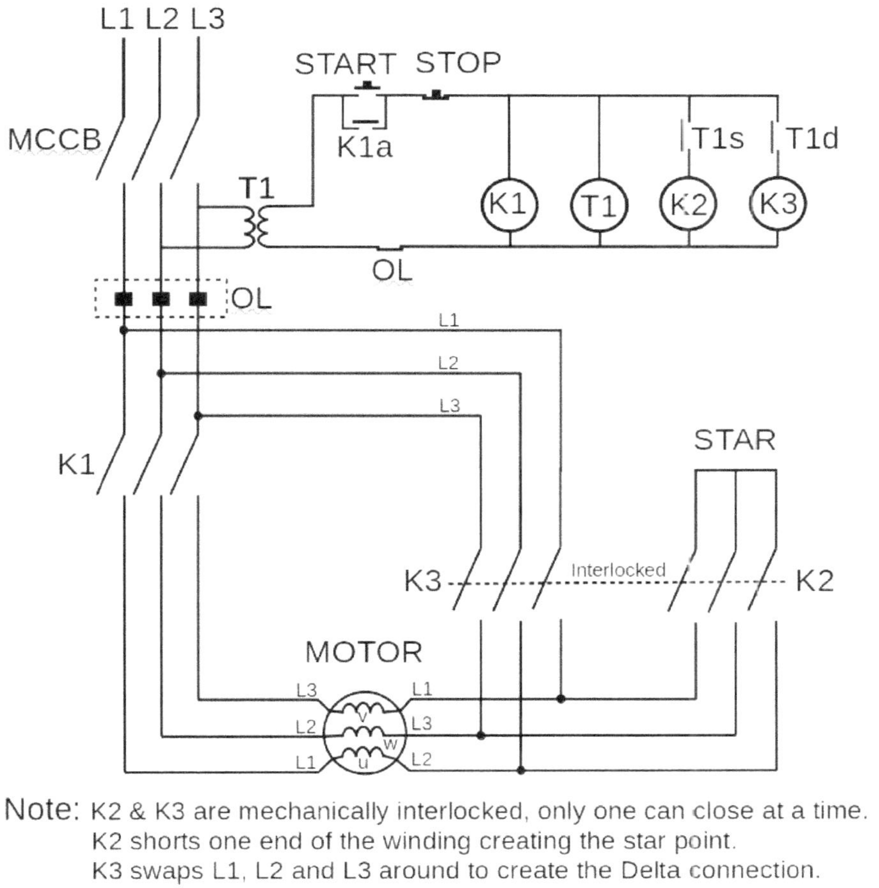

L1 L2 L3
START STOP
MCCB
K1a
T1s
T1d
T1
K1
T1
K2
K3
OL
OL
L1
L2
L3
STAR
K1
Interlocked
K3
K2
MOTOR
L3
L1
L2
L3
L1
L2
V
W
U
Note: K2 & K3 are mechanically interlocked, only one can close at a time.
K2 shorts one end of the winding creating the star point.
K3 swaps L1, L2 and L3 around to create the Delta connection.

Figure 3: *Sample star-delta sketch suitable to answer the above question.*

L1
440V
440/√3
L2
L3
=
Star
K2 RELAY
L1
L2
L3
L1
440V
440V
L2
L3
=
Delta
K3 RELAY
L1
L2
L3
L2
L3
L1

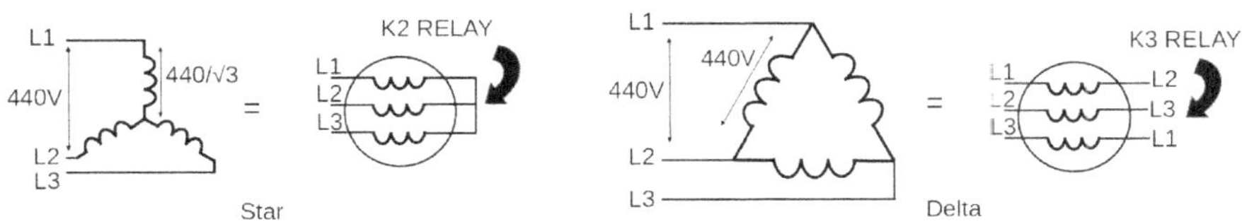

Figure 4: *By closing K2 or K3 with can change from star to delta configuration. We can see the reduction in voltage across the phase coils causing the drop in current and power in star configuration.*

Question 20.

Can you describe the Direct Online Starter (DOL)?

Answer:

- You draw a sketch something similar to that shown in figure 5 and describe it thus:

- When the MCCB is closed, power is available for the control circuit via the step down transformer and to the motor via K1.

- When the START button is pressed, power can flow to contactor coil K1.

- K1 energises, allowing power to the motor causing it to spin up.

- K1a is also closed with K1, shorting out the START button, allowing us to remove our finger from the START button and the motor remains running.

- Pressing the STOP button or if the overload trip operates, the control circuit will be interrupted, causing K1 coil to de-energise and K1 and K1a contacts to open, removing power from the motor.

- Other sensors could be fitted between K1 and motor such as a single phasing detector, a temperature sensor all connected to normally closed contacts next to the OL contact. No matter which sensor detects a fault, the motor will stop.

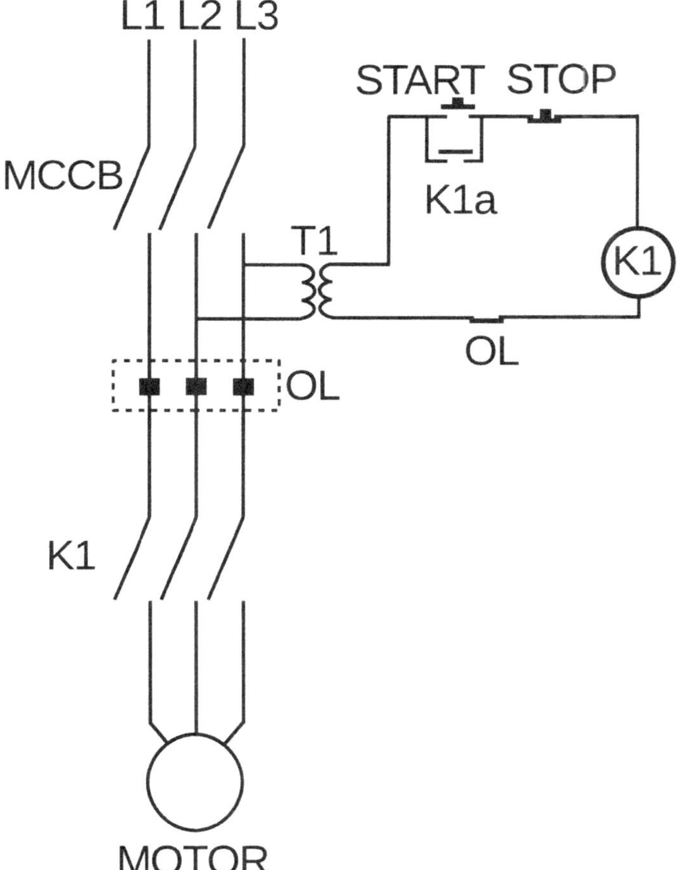

Figure 5: *Sample Direct Online Starter sketch suitable to answer the above question.*

Question 21.

What is wrong with allowing a star-delta motor staying in star mode?

Answer:

- A three phase motor connected to a star-delta starter panel, left running in star mode will consume a third of the current and produce one third of the power compared to delta mode.

- Due to the lower power produced from the motor, it has a greater risk stalling.

Question 22.

If a motor takes a very high starting current, why do the overload trips not open?

Answer:

- Motor overload trips have a delay of several seconds built into them, preventing them from tripping due to the normal current surge during starting.

- If the overload current persists for more than a few seconds, the overload trip activates, shutting down the motor and protecting it.

Question 23.

What are the main types of electrical generators used onboard?

Answer:

- Static exciter and brushless exciter generators.

Blank page to allow next question and answer to be facing.

Question 24.

Describe the brushless generator system.

Answer:

- The brushless generator consists of the main alternator and an excitation alternator sharing a common shaft.

- The excitation alternator consists of stator windings connected to the output of the AVR.

- The AVR supplies DC to the exciter stator windings creating a magnetic field.

- An excitation rotor is within the exciter magnetic field.

- The prime mover rotates the shaft causing the exciter rotor to cut the excitation magnetic field.

- The voltage induced is rectified by the rotating diode pack.

- The DC from the diode pack is fed into the main alternator rotor creating a rotating magnetic field.

- This rotating magnetic field cuts the main alternator stator windings creating the 3 phase voltage output.

- The output from the stator supplies power to the bus bars and the AVR with a Measured Value (MV).

- The AVR is fed with a Set Point (SP) e.g. 440, 690 or 6600 etc.

- The AVR compares the MV to the SP and changes the current through the exciter stator winding to achieve the required voltage.

- The advantage of this type of generator is its decreased maintenance requirements. Absence of carbon brushes increases its reliability.

- The sketch in figure 6 is suitable to answer this question.

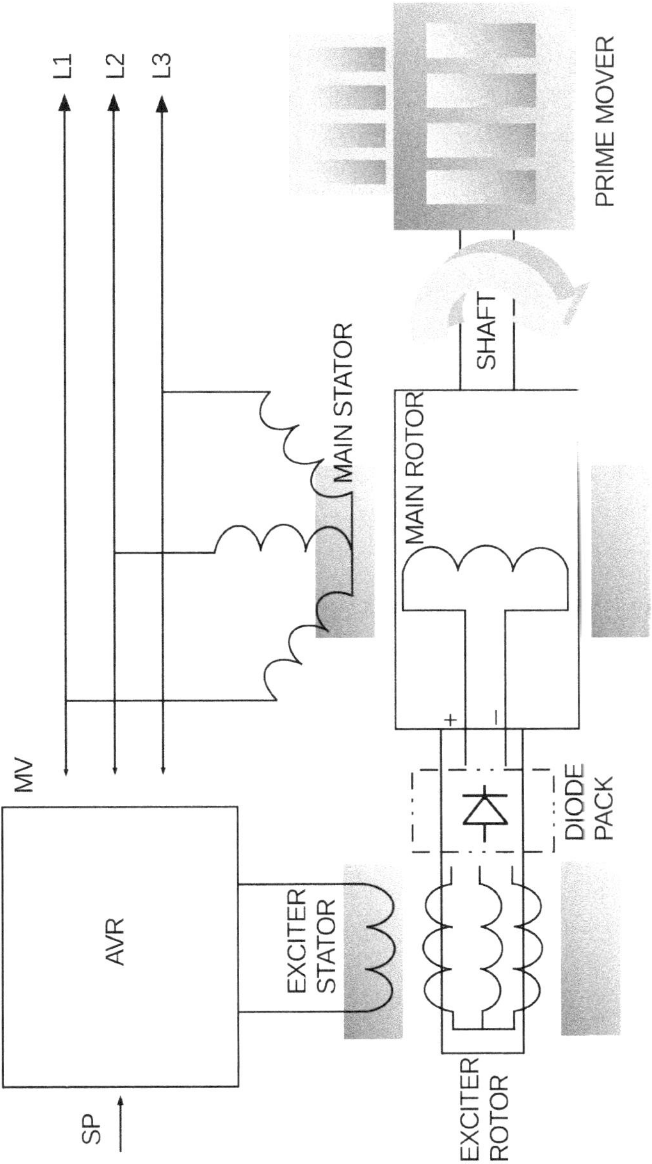

Figure 6: *Sample Brushless Alternator.*

Question 25.

You describe a very basic brushless generator, what main problem do we have with this type of generator in the marine environment and how do we get round this issue?

Answer:

- When a fault occurs causing a large current to flow from the switchboard, the output voltage from the brushless generator drops as its exciter is powered by the generator output.

- This can extend the time it takes for a breaker to trip.

- One approach is to design the exciter to use a permanent magnet. This ensures that as long as the generator is spinning, sufficient excitation voltage and current will be available under fault conditions.

- If the vessel has multiple large loads that switch frequently then a permanent magnet exciter or even a permanent magnet generator can be specified.

- This will reduce the number and size of voltage dips and allow the generator to supply sufficient current while a breaker has time to trip under fault conditions.

Question 26.

You have one generator providing power to the ship via the main switchboard, how do you go about bringing a second generator online, sharing the load at 50% each.

Answer:

- Ask the officer of the watch if the second generator is available and ready to be started.

- Start the generator, check its pressures and temperatures.

- Ensure the generator voltage is within +/- 5% of the bus voltage.

- Adjust the speed of the incoming generator to give a frequency of slightly more than the bus frequency. (typically up to 2 Hz higher)

- Switch on the synchroscope switched to show the difference between the bus and incoming generator phase difference.

- Adjust the incoming speed until the synchroscope is rotating slowly clockwise.

- At 5 to 12, press the close button on this generators breaker.

- Once the incoming generator is providing some load,

- Increase the speed/frequency/governor setting of the incoming generator and, simultaneously:

- Reducing the speed/frequency/governor of the existing generator until both generators are providing 50% of the load.

Question 27.

Why did you arrange for the incoming generator to turn the synchroscope slowly clockwise?

Answer:

- The synchroscope turning clockwise shows that the incoming generator is running slightly faster than the bus, causing the incoming generator to provide power the instant the breaker closes.

- If the synchroscope is turning anti-clockwise, it would mean the generator is running slightly slower than the bus and will take reverse power the instant it is synchronised.

- We set the synchroscope speed to be slow so that when we press the close button at 5 to 12, by the time the mechanism has operated, both generators are in phase at 12.

Question 28.

Apart from the open button, what can can cause the generator breaker to open?

Answer:

- The generator breaker will open if one of the trip relays operate. These monitor: over volts, under volts, under or over frequency, overload, reverse power or phase imbalance.

Question 29.

How do you test these trips?

Answer:

- The reverse power trip can be tested by reducing the power supplied by a generator until the reverse power trip activates.

- The other trips, low/high frequency, low/high voltage and over current etc., can be tested when the breaker is racked out.

- Test voltages and currents can be injected to each individual trip to test trip levels.

- A calibrated test box is used to supply the voltages and currents.

Next question and answer to be facing.

Question 30.

How are these trips connected to the bus bars?

Related question: What are instrumentation transformers

Answer:

- As shown in figure 7, a voltage transformer, connected between two phases, steps down the bus bar voltage to, say, 0 to 10V.

- This reduced voltage is fed into the trip which is set to, say, 6.8V and 6.4V (6.6kV).

- If the transformer has a step down ratio of 1000:1 then these limits equal 6.8kV and 6.4kV.

- Bracketing 6.6kV.

- The under and over frequency detector/trip can be connected to the same transformer.

- The over current trip uses a current transformer shown in figure 8.

- This consists of a secondary winding wrapped round one the bus bars.

- A current transformer can be wound to give, say, 1A output per 10A input, so 100A in, through the bus bar, gives 10A out.

- This current is fed to the current trip with a Inverse Time Relay curve to give us a degree of tolerance of small overloads.

- The inverse time curve allows for a longer period of small overload before tripping and a shorter trip time for a higher overload.

- *WARNING:* A current transformer with an open circuit secondary (or the current meter or trip input is disconnected), acts as a step up transformer, generating very dangerous voltages.

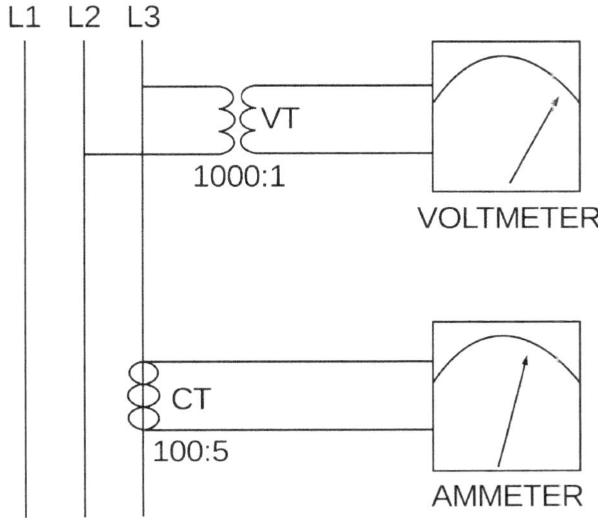

Figure 7: *Instrumentation Transformers with Meters.*

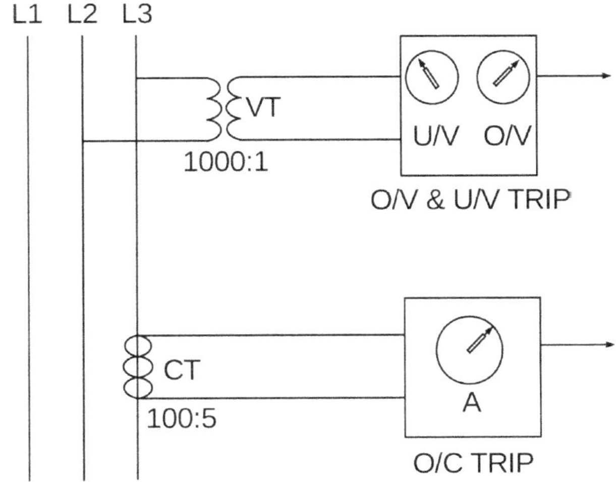

Figure 8: *Instrumentation Transformers with Trips.*

Question 31.

You mentioned the inverse time relay, can you draw me the graph of its response?

Answer:

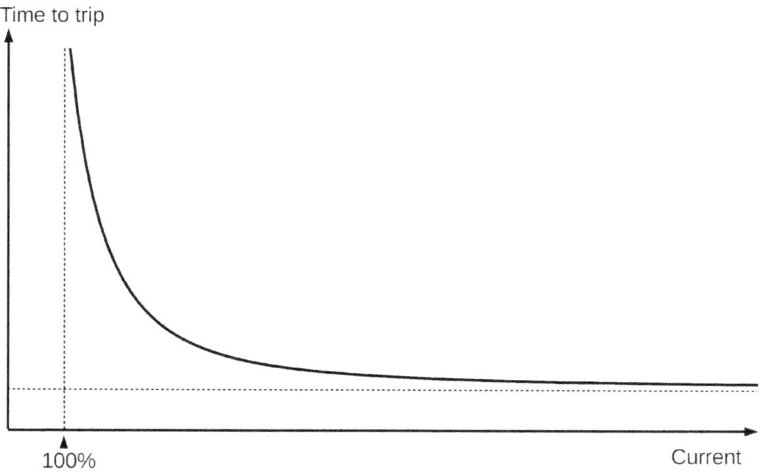

Figure 9: *Inverse time curve showing the higher the overload current, the faster the trip.*

- Figure 9 shows that if the generator is supplying less than 100% power it will continue indefinitely.

- If the generator supplies a little more than 100% power, it will trip after an extended period of time.

- If the generator supplies much more than 100% power, it will trip quite quickly.

Question 32.

Why do some vessels use HV and some LV?

Answer:

- High voltage distribution allows lower current flow therefore much small cable sizes.

- Smaller alternator and electric motor size for same power rating.

- HV is used for a few, very large loads such as bow thrusters.

- LV is used when the supply is required to be used by smaller and more common loads and where human access is more common.

Question 33.

Explain how the governor on a diesel engined generator works?

Answer:

- The governor on a diesel engine has two inputs, the set point (the RPM required).

- The RPM relates directly to the required frequency of the AC output required.

- The other input is the measured value of the RPM of the diesel engine.

- The output of the governor is the fuel rack command signal.

- This controls how far the fuel rack moves, changing the amount of fuel injected into the diesel engine, therefore, its power output and speed.

- Lets assume the generator requires an RPM of 1800 to give an electrical output frequency of 60Hz.

- The set point is 1800, the measured RPM is 1800 and the governor keeps its output to the fuel rack steady.

- When the load on the generator increases, causing the diesel engine to slow down, and the frequency to fall,

- The governor, comparing the 1800 set point with the new measured RPM, will increase the fuel rack command, injecting more fuel, causing the generator to increase its frequency, reaching close to the original 60Hz output.

Question 34.

You said the frequency of the heavily loaded generator is not 60Hz, why?

Answer:

- Droop.

- The governor is designed NOT to keep the generator running at precisely the desired frequency of 60Hz.

- This is because we parallel generators on board.

- With Droop, when the load on a generator increases, the RPM and frequency falls slightly or droops.

- If all generators within an installation have the same droop characteristic set into each of their governors, then when two or more generators, in parallel, sharing the load, have to provide an increase in power, they will all slow down slightly. The generator that tries to supply more power will droop more, causing it to share with the other generators.

- Droop control allow unattended sharing of generators with no computer involvement. Increase the load, and all paralleled generators slow slightly, retaining their load share.

- A droop figure of 4% is typical.

- Without droop, two of more generators in parallel would *fight* to give the most output. Ultimately, causing a blackout.

Question 35.

Is there any other kind of droop?

Answer:

- Voltage droop.

- The voltage output from the generator will fall when supplying a greater load due to I^2R losses.

- The AVR increases the voltage back to near its set point but with a slight reduction in voltage or droop.

- This will keep the voltage stable and ensure the kVAr share remains constant between all paralleled generators.

- With voltage droop, the interaction between generators is kept stable.

- With both speed droop and voltage droop, paralleled generators will share both power and kVAr without intervention.

Question 36.

How do you adjust both the governor and AVR on paralleled generators?

Answer:

- With both generators on the board and supplying power, I would alter the fuel/speed to adjust the power shared between generators and alter the voltage/AVR to adjust the kVAr share.

- All four controls interact with each other and I would adjust them until I had a reasonable power share and kVAr share with the required frequency.

Question 37.

Describe the two types of earthing systems used on board.

Answer:

- On Low Voltage vessels, it is usual to have an isolated earth system.

- This type of system will not trip with one earth fault, allowing the vessel equipment to continue functioning.

- HV systems are *earthed neutral* where the star point of the generator is connected to the vessel hull via a current limiting resistor.

Question 38.

What happens with an earth fault on a HV system?

Answer:

- Earth fault current flows through the neutral earth resistor, which is monitored with a trip, and trips the HV breaker.

Question 39.

Why do we trip the supply, surely we lose the power to the rest of the ship?

Answer:

- The current, power and inherent danger from an earth fault on a HV system means it is safest to shut it down.

- However, the HV cabling is usually short, well installed and protected and only supplies a limited number of large loads, so these faults are very infrequent.

- HV systems are neutral earthed so any single earth fault will always trip the breaker.

Question 40.

Why do most HV vessels have their HV neutral earthed?

Answer:

- Insulated earthed systems can give very high voltage transients during fault conditions.

- A distribution system must have insulation than can cope with the highest voltage expected.

- An insulated HV system transient worst case can be up to 6 times the nominal voltage.

- Classification societies require double the worst case to be applied during vessel commissioning.

- This high voltage necessitates insulation which is prohibitively expensive to install and test.

- Neutrals are earthed via a resistor on HV systems, which limits voltage transients to 2.5 times the nominal voltage and allows for tripping the breaker during any fault condition.

Question 41.

Can you explain how earth faults are detected?

Answer:

- Simple methods include three lamps wired from each busbar through resistor, a switch then to earth.

- If there is no earth fault, all three lamps will glow the same amount.

- If there is an earth fault on one busbar, the lamp associated with that busbar will be dimmer than the others.

- Without the fault present, current will circulate as normal three phase flows into and out of each lamp.

- With the fault present, then the fault will be in parallel with the matching lamp, causing the lamp to be slightly dimmer. The other two lamps will be taking slightly more current and glow slightly brighter.

- See Figure 10 for earth lamp operation.

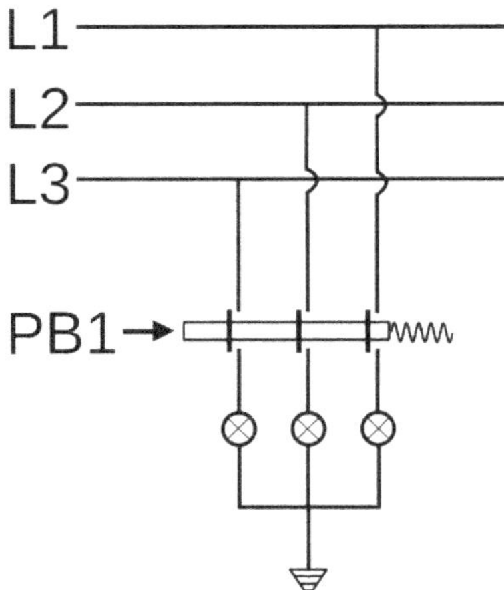

Figure 10: *Three earth lamps.*

Question 42.

Isn't there a more complex system in use today that is more sensitive to earth faults?

Answer:

- There is a more complex, sensitive and more modern way of detecting earth faults. DC current is injected into the busbars and we measure how much in any, current flows.

- With no earth fault present, placing 20V DC onto a busbar has no effect due to there being no complete path for the current to return to its power supply.

- If there is a slight earth fault, then the 20V DC voltage will cause a small DC current to flow through the busbar and monitoring circuit which can be measured.

- Figure 11 shows a sample current injection circuit.

- V1 is the 20v DC power supply, completely isolated from everything onboard the vessel except for two wires.

- The top wire connects to the busbar via R1. The bottom wire connects to the hull of the vessel (Earth).

- If there is NO EARTH fault, then current will not flow from the top of the battery, through R1 then to the bottom of the battery because the busbar has no earth fault, no earth connection.

- Even though the busbar has 660V AC on it, lots of power flowing along it, it has no effect on the circuit. Zero current flows through R1.

- This means no voltage can be measured across R1 so the operational amplifier (+ and -) has nothing to amplify and a very low voltage is output from the opamp, to a meter or an alarm etc.

- However, when an earth fault is present, even though there is still 660V AC on the busbar, lots of power flowing, this small earth fault gives our small 20V power supply a chance to send its 20V out to the busbar, through the earth fault, through the hull of the ship and back to the 20V DC supply. Now the 20V supply can see a clear path and a small current flows through R1 causing a small voltage be measurable across R1. This small voltage is amplified by the opamp and sent to the meter or a trip etc.

- In a real instrument, this circuit will be triplicated, one for each busbar.

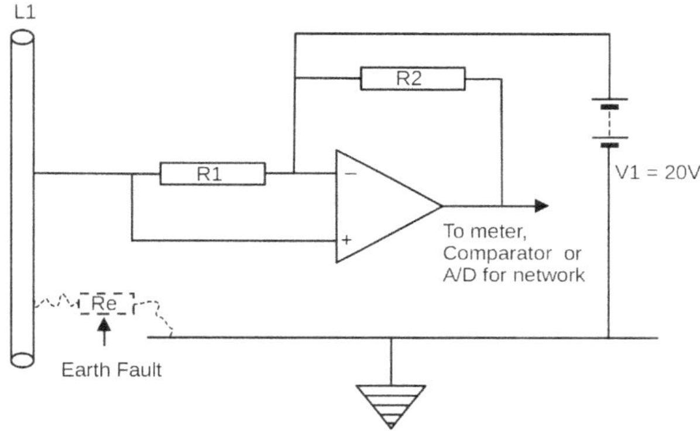

Figure 11: *Detecting earth faults using DC current injection.*

Question 43.

How do we protect the electrical system from faults or overload?
or
How do we maintain supply to most of the ship during a fault or overload?

Answer:

- We use preference tripping and discrimination.

- Preference tripping occurs when a generator is slightly overloaded.

- We automatically shut off non-essential supplies in an orderly way, such as the galley or air conditioning to reduce the load.

- This gives time to put another generator on the board.

- Discrimination occurs when a short occurs in an electrical load.

- The protection device closest to the load will have the fastest and lowest level of trip, causing the faulty load to be isolated.

- Other loads fed from the same supply but higher up (closer to the generator) retain their supply.

- Both preference tripping and discrimination keeps most of the vessel powered most of the time to allow time to fault find and repair.

Question 44.

Do earth faults travel through a three phase transformer?

Answer:

- No, the secondary windings are insulated from the primary winding.

Question 45.

You have been asked to perform some routine electrical maintenance on a electric motor that is driving a seawater pump. It is currently running, it has another motor/pump set on automatic standby. Take me through all of the steps from beginning to end.

Answer:

- I would go to the ships planned maintenance system and print out all of the required documents: permits to work, risk assessments and look up what had been recently done to this motor.

- I would ask the OOW if I can shut the motor off for this maintenance task.

- With their permission, having filled in the permits and assessments, posted the permit in the appropriate places: ECR, Bridge etc. I would lockout and tag the breaker feeding this motor.

- I would proceed to the location of the motor.

- At the motor, guided by the permit, I would lockout and tag the local control of the motor.

- Guided by the permit and risk assessment, I would block off and tag the pump inlet and outlet to prevent the adjacent motor/pump set forcing the motor to move.

- I would inspect the outside case of the motor to check for damage and leaks.

- I would open the terminal cover.

- I would check, using the correct approved measuring device that the motor was dead, using the prove, test, prove technique.

- First proving the measuring device was correctly functioning, then performing the test, then rechecking the measuring device.

- With the motor proven dead on all three phases, I would perform an insulation test.

- I would use the formal testing device provided by the company.

- I would connect to the phases of the motor and motor body, applying the voltage required, as stated in the planned maintenance system.

- I would measure the resistance of each phase and record the values.

- Ensuring the inside of the terminal box was clean and dust free, I would close up the terminal box.

- I would remove the blocks and tags from the pipe work to the motor.

- I would remove the local lockout and tag.

- I would remove the ECR lockout and tag.

- I would inform the OOW that the motor was ready for use after it is proven to work.

- I would input the recorded values into the maintenance system and note that the task was complete.

- I would cancel the permits and close down the paper work.

Question 46.

You mention locking out and tagging, can you explain in more detail please?

Answer:

- Locking out and tagging are a means of safely and securely making electrical equipment safe to work on.

- The locking part is the use of a padlock, which is numbered, and can be used to lock a breaker in the *off* position.

- The tag part is a numbered label that is attached to the lock. The label contains a number/date/time and my name.

- When a lock and tag is fitted, I keep the key in my pocket, preventing anyone else from removing the lock and accidentally closing the breakers causing the equipment under test from becoming live.

- Keeping the locks and tags in a central place enables the system to be safe and easy to use.

- Using two points of isolation and lock out greatly reduces the risk of accidental live working.

Question 47.

What are typical values you would expect for insulation resistance?

Answer:

- I use the formula that most classification societies recommend:

Minimum acceptable resistance

$$= \frac{V}{1000} + 1 \ M\Omega$$

Typical values used are:

440V	=	1 $M\Omega$
1000V	=	2 $M\Omega$
6600V	=	7 $M\Omega$
11000V	=	12 $M\Omega$

but values from SMS will guide us.

Question 48.

How would your maintenance procedure differ if you needed an additional member of staff to help?

Answer:

- I would conduct a toolbox talk to ensure that everyone understood the task, process and their role within it.

Question 49.

What causes earth faults?

Answer:

Breakdown of insulation between a live conductor and the vessels earth. Typically by:

- Water ingress into a light fitting, an electric motor or a junction box.

- Poor connections within an electrical junction board or distribution panel.

- Contamination within an liquid cooled transformer.

- Insulation breaking down within an electric cable.

- Insulation breaking down within an electric motor.

Question 50.

How would you diagnose/find an earth fault?

Answer:

- Take full notice and action SMS procedures and notify affected parties,

- Using the ships distribution layout drawings: I would isolate at the main switch board, distribution panels and other isolation switches and breakers, noting when the earth detection indicator returned to normal.

- I would work towards the fault, isolating then reconnecting power until I approached the furthest from the main switchboard, closest to the faulty load or panel.

- Once the faulty item was located I would enter the SMS to start the step by step maintenance procedure on the faulty device.

Question 51.

Do earth faults pass through transformers?

Answer:

- No. Earth faults are contained within an section of cable between two transformers and any loads in between.

- The earthing effect can not pass through a transformer because the secondary winding of the transformer is completely insulated from its primary winding.

- There is no direct DC path for conduction to take place.

- However, an autotransformer does have a copper path between the primary and secondary windings and will allow an earth fault to propagate through.

- Transformers can have internal earth faults to ground. That is: windings connecting to earth, which would be measurable on the cables connecting the transformer.

Question 52.

What safety devices would the prime mover of an electrical generator have?

Answer:

- If the prime mover was a diesel engine, it could have low pressure sensors for the lube oil(s), high temperature sensors for the lube oil, cooling water, exhaust manifold and an overspeed trip.

- These sensors can connect to alarms and shut down trips.

Question 53.

What's the difference between an alarm and a trip?

Answer:

- An alarm gives an indication that a required value say lube oil pressure is too low, but not low enough to cause damage.

- A shutdown trip will be set to a lower level, at which it has been determined to shut down the engine before irreparable damage is done.

Question 54.

How would you test alarms and trips?

Answer:

- To test an alarm you remove the temperature sensor from the prime mover and insert it into a temperature calibrator and increase the temperature to measure what the alarm trip point is.

- The trip can be tested in the same manner.

- Pressure alarms and trips can be removed and connected to a pressure calibrator. Varying the pressure and checking the alarm activation value and the trip value.

- The calibrators are required to be sent ashore to be calibrated periodically in a laboratory - typically every 12 to 36 months.

Question 55.

What type of sensor could you find measuring the cooling water temperature in a diesel engine?

Answer:

- A PT100 is accurate and very common.

Question 56.

What type of sensor could you find measuring the exhaust gas temperature in a diesel engine?

Answer:

- A thermocouple. A PT100 does not work above approximately 600 degrees Celsius.

Question 57.

How does a PT100 temperature sensor work?

Answer:

- Platinum is a very stable element with known temperature - resistance characteristic.

- Its chemical symbol is Pt.

- A small length of Platinum wire is cut to give a resistance of 100Ω at 0 degrees Celsius.

- A resistance meter connected across the Platinum wire measures the resistance but is programmed to indicate temperature.

- If the sensor detects a rise in temperature, the PT100 resistance increases, the meter measures this and indicates the increased temperature.

Question 58.

How do we remove the inaccuracies created by the cables in a PT100 system?

Answer:

- In a two wire PT100 system, the cable resistance connecting the sensor would be added to the PT100 resistance and give errors.

- By making a 3 wire or 4 wire connection, the cable resistance can be cancelled out giving a very accurate measurement.

- In a 3 wire PT100 system, the third wire allows the length of cabling to be equally added to the top and bottom half of a bridge, cancelling out any length of cabling.

- In a 4 wire system, a constant current is fed into the PT100 sensor. The voltage dropped across it now depends only upon the resistance. This voltage can be transmitted over long distances with little loss.

Question 59.

Can we use either a 3 wire or 4 wire sensor in a 3 wire or 4 wire PT100 system?

Answer:

- It is always best to repair faulty sensors with the official replacement parts, however, in an emergency:

- A 4 wire sensor can be used to replace a 3 wire sensor by leaving one wire from the sensor open circuit, with no loss of accuracy.

- A 3 wire sensor can be used to replace a 4 wire sensor by inserting a link between the two wires from ONE END of the sensor, to ensure each of the wires in the 4 wire connecting cable are in use. Figure 12 shows both variations.

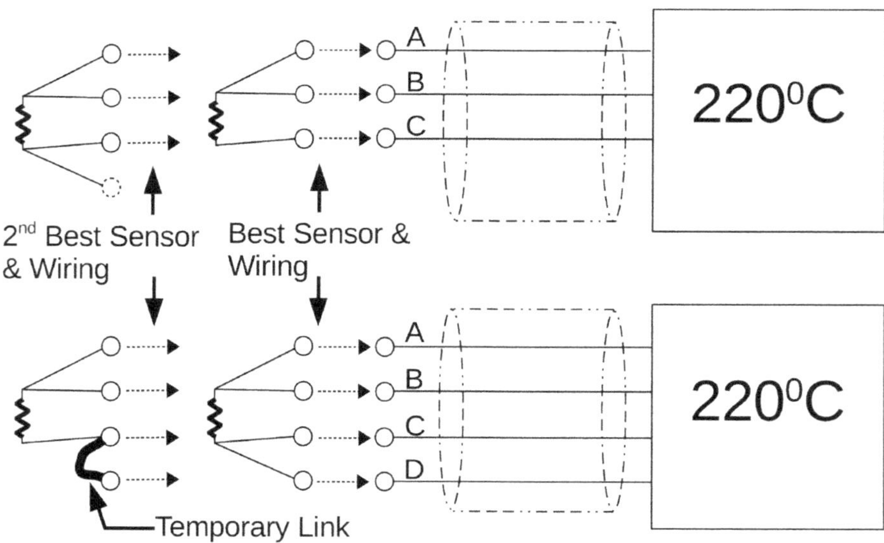

Figure 12: *Wiring for 3 and 4 wire PT100 sensors, normal and get you home mode.*

Question 60.

You mention that HV and LV use two different earthing schemes, what do they do shore side?

Answer:

- Shore side, safety comes before functionality, so one line of a phase pair is connected to earth (Earthed neutral system), which means *any* earth fault will cause a fuse to blow or a breaker to operate. At sea, the same system might trip at a critical time.

Question 61.

Describe a synchronous motor?

Answer:

- A synchronous motor has a electrically energised rotor surrounded by a traditional three phase stator.

- The rotor is excited by DC which causes its windings to produce a large magnetic field.

- When three phase is applied to the stator, the rotating magnetic field created pulls the rotor round.

- Unfortunately the rotor is not self starting and needs the help of an additional motor to get it close to synchronous speed.

- The DC for the rotor can be provided by slip rings with lots of maintenance issues or a brushless exciter can be used similar to a brushless alternator.

Question 62.

Can you explain the starting methods for a synchronous motor?

Answer:

- Originally a synchronous motor was started by a small external motor, sometimes called a donkey motor or pony motor.

- Modern synchronous motors have a starter motor coexisting on the same shaft using windings built into the rotor.

- Effectively, the starter motor is an induction squirrel cage within the synchronous rotor.

- Three phase is applied to the stator winding, the built-in squirrel windings work as expected and the motor runs up to 95% of synchronous speed.

- The control circuit allows DC current flow through rotor, causing the rotor speed to increase to synchronous speed.

- While the motor runs at synchronous speed, no current is induced into the squirrel cage.

Question 63.

Why do we use synchronous motors?

Answer:

- A good use is power factor correction.

- If the rotor is fed with more DC current than is necessary for the motor to run correctly, the three phase stator terminals start to present a *leading* power factor to the three phase supply.

- The higher the DC current fed into the rotor, the greater the *leading* power factor.

- Synchronous motors are the most efficient type of high power motor.

- Shoreside, Synchronous motors without a load, are used for large scale power factor correction and are called *Synchronous Condensers*.

Question 64.

Answer:

- Due to the loads on a typical vessel (induction motors), the power factor on the board is less than 0.8 lagging.

- This creates extra cost in running the generators because we have to supply both the *real power* and *reactive power*.

- By running the synchronous motor with a leading power factor cancels some of the lagging power factor.

- We typically run with a power factor of 0.8 - 0.85

- We can also use a bank of capacitors across the three phase supply to cancel out some of the lagging power factor.

Question 65.

Why do you not run with a power factor of 1.0 as that would be most efficient?

Answer:

- All of the generator trips and power control systems work correctly only with a lagging power factor.

- On a vessel, the amount of load is changing throughout the day with motors being started and stopped.

- If the power factor was corrected to be 0.99 lagging, then switching one motor off could give you a leading power factor, causing instability and trips to occur.

Question 66.

What bridge equipment does the GPS feed?

Answer:

- The GPS sends location (latitude, longitude and speed) data to the ECDIS, Radar, VDR and Gyro.

Question 67.

Why does the GPS send data to the Gyro?

Answer:

- For best accuracy the Gyro requires the vessels latitude (north or south of the equator) and speed, to correct Gyro errors.

Question 68.

How can a GPS be made more accurate?

Answer:

- DGPS could have been used.

- Differential GPS makes use of nearby, fixed land stations (DGPS Beacon stations) that receive the same satellite signal as you. If the ships position via GPS is incorrect, so will be the fixed land stations position. This error difference between the land station actual position and its GPS measured position is transmitted on the Medium Frequency band (approx 300kHz) to all ships within range (approx 300 nautical miles). Our ship receives these error signals, applies them to our position and produces a more accurate result.

- DGPS has been discontinued due to steady improvements in GPS receivers over the years.

- Satellite Based Augmentation Systems (SBAS) is a satellite based differential correction system which is slowly spreading worldwide. Some states still use a version of DGPS but users must ensure they are using an approved signal source.

Question 69.

The deck OOW calls you and says the radar appears not to be working, what would you do?

Answer:

- I would confirm with the OOW what the symptoms are.

- I would go through the user setup of the radar (gain, range, clutter etc.) to ensure that their was no maladjustment.

- I would check the tuning as it is normally left on automatic tune.

- I would turn on the performance monitor and compare the monitor result to the last recorded performance test.

- The performance monitor result would indicate either a transmitter fault (magnetron or modulator) or receiver fault (TR cell, mixer diode or local oscillator).

- I would run any signal processing diagnostics.

- Due to the lack of spares and the inaccessibility of the radar turning unit which usually contains the transmitter receiver, I would recommend we arrange for a service technician to come on board on arrival in our next port of call.

Question 70.

The deck OOW calls you and says the GPS position information is missing on the Radar display, what would you do?

Answer:

- I would confirm with the OOW what the symptoms are.

- I would check that the GPS is powered up and other users of GPS data are working correctly.

- If all users of GPS data were working except for the radar I would check the NMEA GPS input to the radar.

- If this was okay, then the radar unit is at fault. fit a spare or arrange for a shoreside technician to do so at the next port.

- If there was no data at the input to the radar I would check the output of the GPS that feeds the radar.

- If this was okay I would check and repair the cable.

Question 71.

Tell me some ways of measuring liquid levels in tanks?

Answer:

- We can measure liquid levels in a number of ways:

- Float.

- Hydrostatic pressure.

- Reflecting sound or radio waves off the surface.

- Capacitive probes.

Question 72.

Explain a level sensor that uses hydrostatic pressure?

Answer:

- Fit a pipe to the top and bottom of a tank containing liquid.

- Connect the pipes to a differential pressure cell (DP Cell).

- As the liquid level rises, the pressure difference between the bottom of the tank with respect to the top of the liquid will increase.

- The DP Cell will measure this pressure difference and transmit a signal directly proportional to the level of liquid within the tank.

Question 73.

Will this DP Cell system work if the tank is either open or the top is sealed?

Answer:

- It will work under both conditions.

- The pressure in a sealed tank can increase but this increase will affect both the top and bottom of the tank equally. The DP Cell will only respond to the difference in pressure due to the head of liquid.

Question 74.

What type of fire detection sensors are you aware of?

Answer:

- Smoke, flame and heat.

Question 75.

Why do we have different types of fire sensors?

Answer:

- We want to maximise the chance of detecting a fire quickly and minimise false alarms.

- In a galley with frying and other smoke generating activities we could fit flame and temperature rise detectors.

Question 76.

Describe a common method of connecting these sensor heads to a fire control panel?

Answer:

- The most common method is to use a loop or loops connecting many sensor heads to the fire control panel.

- This allows for a single break in the cabling but the sensor heads still have one signal path back to the control panel.

Question 77.

If the heads are connected in a loop how do you know which head has activated?

Answer:

- Each head is addressable and sends a unique signal back to the alarm panel allowing us to identify the head and compartment at risk.

Question 78.

How do you test fire sensor heads?

Answer:

- The planned maintenance system will instruct us to test a number of sensor heads per day or per week depending upon the size of the vessel. It will provide or point to the instructions for doing so.

- After liaising with the ECR and bridge, manning the fire control panel, in radio contact with the test team and panel team, switching the alarm system to standby for the loop under test:

- Each sensor head on the list would be tested with application of smoke, heat or flickering light to trigger the sensor, and ensure that the alarm system detected it.

- After a number of heads have been tested, the alarm system is brought fully online, the bridge and ECR informed and the planned maintenance system updated.

Question 79.

Do fire sensor heads ever fail?

Answer:

- Yes, they get dirty or connections vibrate loose and give intermittent alarms. Occasionally they fail to respond to the test stimulus.

Question 80.

What is an LVDT and where could we use one?

Answer:

- A Linear variable differential transformer (LVDT) is a sensor.

- It senses movement.

- It is constructed using windings that are spread out along a former, one primary and two secondary windings.

- The primary winding, located inline, between the two secondary windings is energised with a high frequency voltage.

- This primary voltage induces a small voltage into both secondary windings.

- The size of induced voltage depends upon a movable rod which is within both the primary and secondary windings.

- With a suitable choice of rod material and windings, the device is very sensitive to movement of the rod.

- The output of the secondary windings is processed to give a direct voltage which alters in a linear way with small movements of the rod.

- A classic use for an LVDT is a fuel rack position sensor.

- The fuel rack position can be measured to within a fraction of a millimetre.

Question 81.

What is a capacitor and where and how could we use one on a vessel?

Answer:

- A capacitor consists of two conductors separated by an insulator. To get maximum capacitance in a small volume we use metal plates placed very close to each other.

- Capacitors store electrical charge, block the path of DC and pass AC.

- We use lots of capacitors in electronic equipment but in the engine room, we use them for power applications:

- We use them for power factor correction, they give a leading PF.

- Starting capacitors for single phase motors.

- In a DC link we use one to smooth pulsating DC to smoother DC.

- In variable speed drives we use them to reduce (filter out) harmonics.

- We use capacitance created by fitting an insulated tube into a cargo tank. The capacitance created between the metal tube and the ships hull gives a measure of the liquid in the tank. The capacitance value changes due to the dielectric changing between the tube and the tank sides when the cargo level rises or falls.

Question 82.

You mention harmonics, what are they?

Answer:

- In a vessels power system, harmonics are unwanted higher frequency voltages that can disrupt some electrical equipment.

- When a motor is turned on of off, there is a surge of current into the motor.

- This surge generates voltage pulses (harmonics) which can interfere with electrical systems.

- Variable speed drives can generate large amounts of harmonics due to their controllers taking large, short, sharp, intermittent amounts of *current* from the ships supply.

- These harmonics can be filtered out to some extent but the filtering equipment is another item of equipment that could fail and does require regular maintenance.

- Harmonics cause excess heating in generators, motors and transformers.

Question 83.

If you had exceptionally sensitive equipment onboard (scientific?) How could you ensure an ultra smooth supply?

Answer:

- On vessels that require ultra noise free (harmonic free) supplies, it is common to use a genset or a Motor Generator set.

- This is an AC induction motor powered by the normal dirty three phase, mechanically connected via a shaft to a three phase voltage generator. The generator has zero connections to the main ships supply and the output of the three phase generator remains smooth and clean with zero harmonics.

Question 84.

What are phase shift transformers?

Answer:

- Phase shift transformers insert a phase difference between their input and output connections of up to +/-35°.

- Their most useful characteristic is the ability to cancel out odd harmonics generated by large electrical loads such as variable speed drives.

- Phase shift transformers can be designed to cancel out or filter out a number of odd harmonic frequencies and prevent these harmonics from travelling along the power distribution system.

- Internally, phase shift transformers create a 90° phase shift between the voltage on the primary winding of the excitation transformer compared to the voltage on its secondary winding.

- They typically consist of two separate three phase transformers, the first connected across the supply, the second transformer connected in series with the supply lines.

- The secondary windings of the first transformer are connected to the primary windings of the second transformer.

- Depending upon the turns ratio of the excitation transformer, a controlled amount of phase shifted voltage is applied to the primary of the series transformer, combining the in-phase, incoming voltage with the 90° phase shifted voltage to give an overall phase shift.

- It is common to have a phase shift transformer installed between the main supply and each variable speed drive.

- Some variable speed drives contain a phase shift transformer as a means of creating a 30° phase shift used to trigger the thyristors within the VSD.

- Some phase shift transformers contain taps on the first transformer which allow different phase shifts to be selected, which allows different harmonics to be selected for cancellation.

- Some PST's can reverse the phase of the first transformer secondary which allows a +30° phase shift become -30° phase shift.

- Maximum cancellation occurs when each phase of the three phase supply is balanced.

- Even harmonics on a well balanced three phase supply automatically cancel out. Figure 13 shows a typical layout.

- A simple transformer with a delta primary and a star secondary will give a 30 degree phase shift.

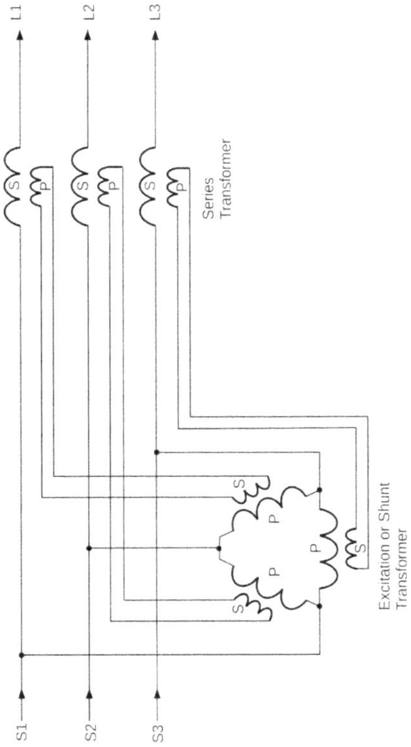

Figure 13: *Phase shift transformer showing shunt and series transformers.*

Question 85.

How does a main circuit breaker work, what's inside?

Answer:

- A main circuit breaker will be connected between each generator and the bus bars.

- It contains moving contacts that make and break, connecting the generator output to the bus bars.

- The current flow through the breaker contacts, when the breaker opens, is so high that the energy within the arc can be damaging. A variety of techniques are used to minimise or control the arc.

- On low voltage systems (LV) the arc is sufficiently small, using high speed moving contacts, air can dowse the arc quickly. Guiding and cooling the arc with an arc chute can prevent excessive wear of the contacts.

- The moving contact is energised by a strong spring, so the contact can be *tripped* even during black out conditions.

- The spring is automatically retensioned by an electric motor.

- The spring can be retensioned manually if the motor has failed.

- High Voltage breakers, due to the increased power available and the strength of the arc, one of two methods can be used:
  - Vacuum breaker
  - SF6 breaker

- Vacuum breakers keep two contacts per phase in a vacuum chamber. An arc has great difficulty travelling through a vacuum. So much so, the distance between each phase contact is very small (less than 10mm).

- Sulphur Hexafluoride (SF6) greatly inhibits arc travel and enables a small gap to be used. Two contacts per phase are contained in a sealed chamber of SF6. The chamber is pressurised with SF6 to around 3 to 7 bar.

- A pressure gauge indicates the SF6 pressure.

- The construction of vacuum and SF6 breakers are similar.

- A bellows allows the moving contact to touch the fixed contact without impairing the integrity of the gas seal or vacuum seal.

- The contacts are shaped to prevent arcing from damaging the contact surface.

- New designs are being produced that reduce the wear on contacts, greatly increasing the life of the contacts.

- Due to climate change, installation of new SF6 units will be banned starting in 2026 (EU 2024/573). Existing SF6 installations can still be used, repaired and serviced. At the time of publishing, no new MSN's or MGN's cover this change.

- New gases are being developed and tested to replace SF6.

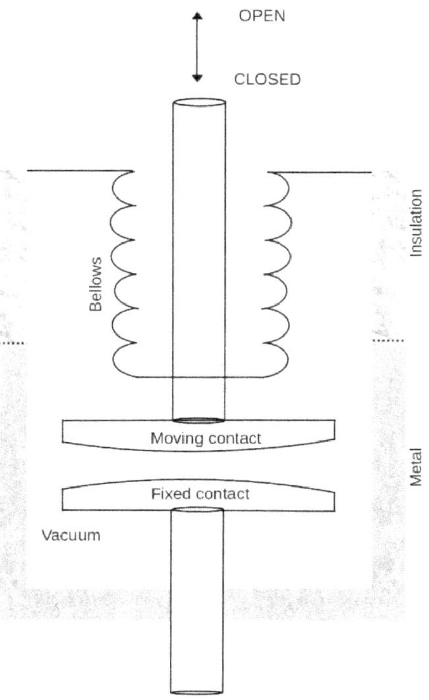

Figure 14: *Vacuum breaker.*

Question 86.

How would you test an HV breaker?

Answer:

- A high voltage pulse generator (approximately 10kV for a 6.6kV breaker) can be used to test the quality of the vacuum in a vacuum breaker. Many shipping companies use shore side technicians for this but you need to be aware of it.

- A vacuum breaker can be verified by using a **VIDAR** instrument from Megger which generates high voltage DC pulses.

- A voltage pulse of a specified value is applied. A breaker with an impaired vacuum will arc at lower voltage levels and with a good vacuum, not arc.

- The VIDAR gives a simple pass / fail result and requires two switches to be operated simultaneously, requiring two handed use, increasing safety. See figure 15

- An SF6 breaker can be tested by measuring the pressure of the SF6 gas.

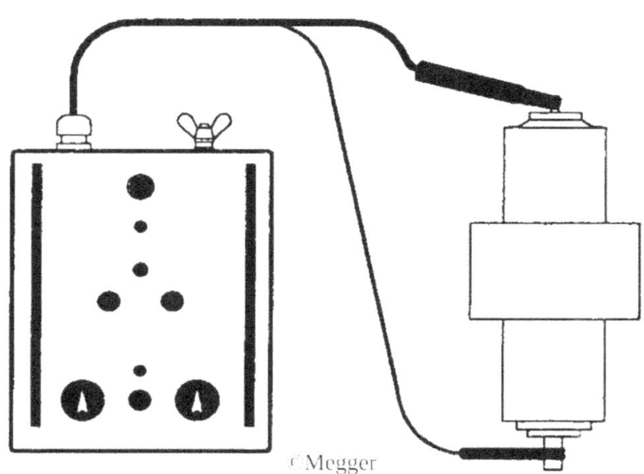

Figure 15: *Vidar tester from Megger connected to Vacuum breaker* ©*Megger.*

Question 87.

Why do some vessels use diesel electric propulsion?

Answer:

- Diesel electric propulsion is used to give more flexibility of thrust.

- By using a variable combination of generators, fuel efficiency can be enhanced during slow speed running.

- If combined with Azipods, efficiency is increased.

Question 88.

Why do azipods increase efficiency?

Answer:

- Azipods are located in *clear water* away from the hull. The propeller is a puller not pusher. These features allow the propeller to grip undisturbed water which increases its efficiency.

- Due the extra manoeuvrability enabled by azipods, tug hire during berthing operations can be minimised, saving additional costs.

Question 89.

What types of diesel electric speed control systems are there?

Answer:

- The most common are:
- Cyclo converter
- Synchro converter

Question 90.

Explain how a cyclo converter works?

Answer:

- A typically cyclo converter changes AC to variable frequency AC.

- It uses thyristors to switch the incoming three phase AC directly to the propulsion motor.

- By controlling the triggering of the thyristors, a variable frequency is output.

- The timing of the thyristor triggers varies to give the required output frequency.

- The output frequency can not exceed about half the input frequency.

- For maximum efficiency, the output frequency should be less than a third of the input frequency (20Hz).

Question 91.

Explain how a synchro converter works?

Answer:

- A synchro converter changes AC to DC then to variable frequency AC.

- Three phase input is rectified to DC and connected to a capacitor where the fluctuating level is smoothed.

- This DC voltage is named the *DC link*.

- The voltage across the DC link capacitor is fed to a bank of three phase thyristors which are switched, in turn, to produce a three phase signal that can be fed to the main propulsion motor.

- Synchro converters can produce frequencies twice as high as the input frequency (120Hz).

Question 92.

Can variable speed electric drives cause problems to the electrical distribution? *or how are harmonics made and how are they reduced?*

Answer:

- The rapid switching of the supply voltage by SCRs, thyristors, IGBTs or other high speed switching devices within variable speed drives, cause high frequency voltage pulses to be superimposed onto the switchboard outputs, interfering with ship board electronics. These pulses corrupt the sine wave output of the generators and are called *Harmonics.*

- Harmonics are reduced by the use of filters (capacitor banks, inductors, transformers) or by use of improved switching technology.

- Some ships, which have a large and sophisticated electronics load, use Motor-Generator sets to produce isolated three phase supply that has no direct connection to the ships distribution system.

- This isolated three phase supply, having no direct connection to the ships supply has negligible harmonic content.

Question 93.

Are harmonics a problem?

Answer:

- Yes - two reasons

- Harmonics can interfere with normal working of electronic equipment.

- Harmonics can increase the amount of power dissipated in generators, cabling, motors and transformers.

Question 94.

You mention IGBT, SCR and others in your harmonics answer, explain them please?

Answer:

- An Insulated Gate Bipolar Transistor (IGBT) is a semiconductor device that consists of a high power bipolar transistor, driven by an insulated gate Field Effect Transistor (FET) in one small package.

- This combines the low voltage input and high input impedance characteristic of a FET, with the high current and low volts drop characteristic of a bipolar device.

- IGBTs can be connected in series to increase the overall level of voltage they can switch, and be connected in parallel to increase the overall current they can switch.

- They commonly come in *stacks* (multiple IGBT devices) that can switch on and off, voltages of 690V at 2000A.

- Multiple *stacks* can be connected to control much high voltages and currents.

- Silicon Control Rectifiers (SCR) or Thyristor, are a diode type of device that do not conduct until a pulse or voltage is applied to its gate input.

- The SCR then operates exactly like a conventional diode.

- By controlling the time of the pulse arriving at the gate of the SCR, you can control how long the SCR conducts before the AC goes negative and the SCR stops conducting.

- SCRs were the first high voltage and high current semiconductor devices created which is why many variable speed AC drives use them.

Question 95.

You mentioned Motor-Generator sets, what can you tell me about these?

Answer:

- A motor generator set is an three phase motor driving a three phase generator.

- Typically sharing a common shaft, the motor spins the generator.

- The motor and generator are electrically isolated.

- The generator will be brushless requiring minimum maintenance and maximum reliability.

- By sharing a common shaft, only two shaft bearings are required.

- The motor is driven by the ships supply which could have many harmonics.

- The generator output will be a good quality sinewave that can be fed to sensitive electronic equipment and systems.

- If the output of the motor generator set does not drive switching power supplies, its output waveform will be nearly harmonic free.

- If the motor part is synchronous, the output will track the main bus bar frequency.

- The generator part can have different number of poles enabling frequency conversion, such as 60Hz to 50Hz.

Question 96.

Describe a proportional controller?

Answer:

- A proportional controller has two inputs and one output.

- Set Point (SP) and Measured Value (MV) are the inputs and Controller Output (CO) is the output.

- The SP has the desired value set.

- A sensor measuring the process variable (temperature, pressure, viscosity, RPM etc.) feeds its value into the MV input.

- The CO is fed to the correcting element, often a pneumatic or electrical valve.

- If there is a difference between the MV and the SP (called the offset), this offset value is multiplied by the gain setting.

- This new value is sent to the CO and hence on to the control valve.

- A well designed system, the CO signal will cause the MV to move towards the SP quickly and stabilise.

- A controller does not know what it is controlling, it just compares the SP with the MV and calculates the CO.

- Increasing the gain reduces the offset but reduces stability.

- A proportional only controller will always have an offset since if offset is zero, the the CO will be zero and the correcting device (valve etc.) will remain closed.

- In figure 16 we can see the proportional controller represented by a box with two inputs and one output.

- On the front of the controller, a gain control can be seen. Within the box, the measured value (MV) a number and the set point (SP) are subtracted from each other. The result is called the offset. The offset is then multiplied by the gain and sent to the controller output. Typically, the CO feeds a valve, pump, heater or cooler.

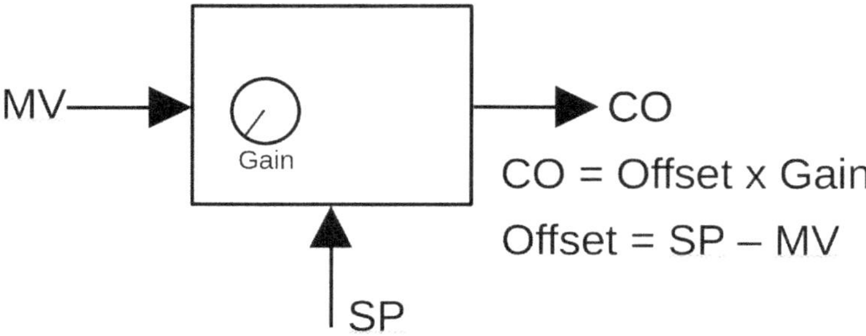

Figure 16: *Proportional Controller.*

Question 97.

What about the other two terms of a PID controller?

Answer:

- The second term on a three term controller is the integral term.

- The integral input integrates the offset. The longer the offset exists and the greater the size of offset, will cause a larger integral signal to be produced.

- The integral signal is added to the proportional signal and fed to the correcting device connected to the CO.

- Use of the integral function causes the correcting element to activate until the offset has reduced to zero.

- The only way for a PID controller to give you zero offset is by using integral action.

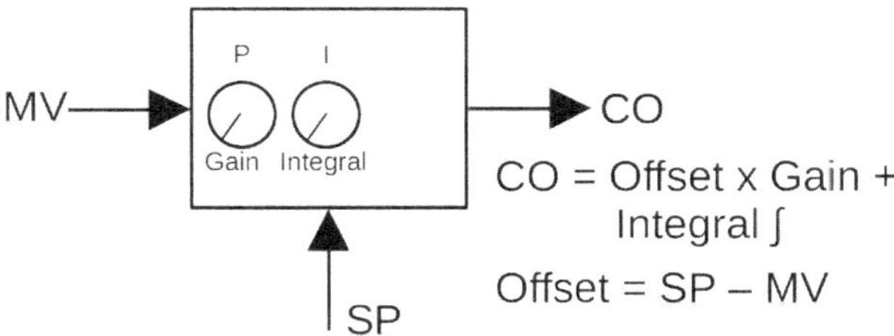

Figure 17: *Proportional plus Integral Controller.*

Question 98.

And the third term?

Answer:

- The third term is the derivative. It causes the controller to act upon rapid changes to the offset.

- If derivative is enabled, a rapidly changing offset will trigger a large and rapid change in the derivative output which is added to the controller output.

- The derivative causes the actuator to respond rapidly to changes in the measured value.

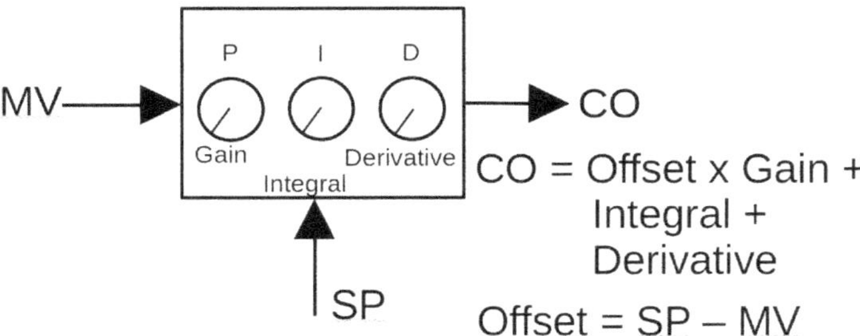

Figure 18: *Proportional plus Integral plus Derivative Controller.*

Question 99.

How do we tune a three term controller?

Answer:

- This is usually undertaken by shore side technicians however, if a control loop required setting up a sea I would

- reduce the gain, integral and derivative actions.

- Increase gain step by step until the measured value starts to oscillate, back off gain slightly until the MV stabilises.

- Increase the integral, step by step, until the offset has been removed.

- Increase integral until the MV oscillates again.

- Reduce the gain slightly until oscillation is just removed.

- Test the system by changing the set point for a few seconds, then returning the set point to its desired value.

- A correctly adjusted control loop tuned using this method should overshoot then undershoot a few times then return to the required value.

Question 100.

You did not adjust the derivative control?

Answer:

- No, manual adjustment of the derivative is not recommended as this can easily give rise to an unstable system.

- I would enable the auto tune function if fitted, or

- use a laptop with a specialist tuning program or

- use the Ziegler and Nichols method which calls for a method similar to the trial and error steps above.

- As per trial and error, get the system just oscillating, record the gain setting and the cycle time for one oscillation.

- Enter these two recorded values into the Ziegler and Nichols formula and extract the gain, integral and derivative values.

- Enter these settings into the controller and test the system response to a step change of the measured value or set point.

Question 101.

You mention testing a newly adjusted control system with a step change, why and how do you do this?

Answer:

- Whenever you adjust a PID control system, there is always a risk that although the system is now working and appears stable, it may become unstable if you have given it too much gain.

- A system with too much gain can suddenly oscillate (tank levels rising and falling, RPM hunting etc.).

- After adjustment, force a step change, e.g. change the set point by 5% for 1 minute, set it back to its original setting and observe what happens.

- If the system is to be declared stable, the set point change will trigger a change to the correcting element (valve) causing the measured value to change.

- After a while (depending upon the physical size of the system) the controller will cause the measured value to oscillate two or three times (typically) then settle back to its original level.

- This is classed as a stable system.

- If you have too much gain in a large system, you may not observe anything untoward until something disturbs the system or you change the set point.

- Change the set point, then reset it back to its normal setting.

- With too much gain, the correcting element will move too far, over correct, then under correct repeatedly.

- The measured value will surge from a peak level to minimum level and back again, indefinitely.

- The method used to stop this oscillation is by reducing the gain slightly.

- The rule is: after adjusting a PID controller - test it with a step change of the set point.

84

Question 102.

List the major electronic navigational aids located on the bridge?

Answer:

- Gyro, Radar, ECDIS, GPS, AIS, Speed Log, Echo sounder, Autopilot, Navtex and Radio communications system (GMDSS).

Question 103.

Describe a gyro and why is it fitted?

Answer:

- Legislation drives its fitting.

- It can consist either a spinning mass or an solid state laser.

- The spinning mass gyro consists of a heavy mass rotated at many thousands of revolutions per minute.

- Due to the spinning inertia created, it *wants* to point to a *fixed* point in space.

- By the addition of a centre of gravity sensor (pendulum) and the application of very small correcting forces creating precession, the spinning mass is made to point to *true north.*

- A laser gyro has a split beam of laser light, one beam passing clockwise through a few kilometres of fibre optic cable and the other beam travelling in an anti-clockwise direction through a separate fibre optic cable.

- Both beams exit the fibre optic cables and land on a single spot.

- If the gyro rotates clockwise or anti-clockwise, the phase of light beams interfere with each other and produce *flashes of light.*

- The number of flashes directly relates to the rotation of the gyro.

- A north seeking laser gyro typically requires three gyros mounted at 90 degrees to each other to capture gyro movement. When the vessel location is input, the gyro calculates the true heading of the vessel.

Question 104.

What is a VDR and what function does it perform?

Answer:

- The Voyage Data Recorder (VDR) is required by legislation.

- It records vital information that can be used to discover the cause(s) of a marine incident.

- It records: Heading, Helm command, Rudder angle, engine command, engine response, RPM, bridge audio, Radar display, AIS data, GPS data, Log, Echo Sounder, Main alarms, Watertight doors, VHF communications, Engine room vital temperatures and pressures.

Question 105.

How often is the VDR tested?

Answer:

- The VDR is required by legislation to have its performance annually tested.

- It must be tested by an approved testing or servicing facility.

- A copy of the compliance certificate must be retained on board.

Question 106.

Explain AIS?

Answer:

- Automatic Information system (AIS) receives operational data from vessels within a 30 mile radius and displays this data on a Radar screen, ECDIS display or AIS display. The data contains the heading and speed of the vessel and other data of interest to the OOW.

- Our ship AIS equipment transmits our heading and speed for other vessels to display and use in a similar manner.

- AIS uses two dedicated VHF channels.

- It contains its own GPS and is fed with the ships main GPS, checking that both locations are in close agreement.

Question 107.

Does the AIS transmit data at a regular or variable time?

Answer:

- The AIS increases its transmission rate when the vessel is changing its heading.

Question 108.

What is NAVTEX?

Answer:

- NAVTEX is a terrestrial, radio based, system that uses a medium wave band frequency (518 kHz) to transmit navigation and safety messages to vessels.

- These messages include navigational warnings, weather warnings, ice reports, search and rescue information and pirate warnings.

- It has an operational range up to 300 miles from land.

- It is required to be carried by legislation.

- Systems used to print out messages but modern systems display messages on a scrollable LCD screen.

Question 109.

Explain what an ECDIS is?

Answer:

- Electronic Chart, Display and Information System (ECDIS) displays navigational charts for use by navigation officers.

- It is a PC based system, that uses preloaded and approved charts, fed with the ships latitude, longitude and heading.

- It allows paper charts to be dispensed with if a dual ECDIS is installed.

- Tankers greater than 3,000 GT and other cargo vessels greater than 10,000 GT are required to carry an ECDIS.

- ECDIS only ships are required to have separate UPS for each ECDIS system to remove a single point of failure.

Question 110.

What is an Autopilot and how does it work?

Answer:

- An autopilot steers the ship on a constant course.

- It is required to be fitted by legislation.

- It receives the ships heading from the Gyroscope and compare this to the required heading which is input by the officer of the watch.

- Any discrepancy between what is required and what the actual value is causes the rudder to move to correct the heading error.

- Modern Autopilots *learn* how much rudder command is required to achieve a heading correction, to minimise the size of rudder movement which reduces fuel consumption of the vessel.

- The Autopilot is effectively a three term controller and we want the ship to keep as close as possible to the required heading with the minimum of rudder action.

Question 111.

What is a Zener barrier and where is it used?

Answer:

- A Zener barrier is a device that prevents excess voltage and current entering a hazardous space.

- It reduces the maximum voltage and current available, ensuring any accidental spark within an enclosure will contain insufficient energy to ignite the gas, vapour, mist or dust in a hazardous space.

Question 112.

Okay, you have described in concept what a Zener Barrier does, how does it do it?

Answer:

- A Zener barrier consists of series resistors and parallel zener diodes.

- If the input voltage rises by more that a certain amount, the zener diode(s) conduct, preventing the higher voltage from reaching the output of the barrier.

- If a short occurs on the output side of the zener barrier, the series resistors limit the maximum current that can flow in the protected space.

Question 113.

A Zener barrier has been damaged, how would you repair it?

Answer:

- Zener barriers are sealed units that are repaired by replacement.

- Some zener barriers contain a fuse, but this is contained within the body of the barrier and not accessible.

- Zener barriers are sealed within a *two part epoxy potting compound* which ensures maximum protection from dangerous external conditions affecting the internal circuitry and, prevents incorrect repairs from being conducted, which could leave the zener barrier in a dangerous state.

Question 114.

Why do we need special enclosures for electrical equipment within hazardous spaces?

Answer:

- We need special enclosures for electrical equipment in hazardous spaces to prevent explosion or fire.

Question 115.

Can you explain more?

Answer:

- Enclosures come in different classes to suit the hazard within a space. For example:

- Intrinsically safe enclosures are used to house electrical and electronic equipment when fitted in a dangerous space.

- A dangerous space includes compartments that have or are expected to have dangerous levels of explosive gas, vapour, mist or dust.

- To simplify equipment requirements, spaces, gases, vapours, mists and dusts are categorised with regards to the level of protection required to achieve safe operation.

- Gases ignite at different surface temperatures or different energy levels from sparks.

- Spaces can occasionally be subject to explosive gas or constantly containing explosive gas.

- Vessel design, class rules and international standards dictate which level of protection is fitted to which area.

- The enclosure is designed to prevent a spark or hot spot from faulty equipment from igniting the gas, vapour, mist or dust within the compartment or near the compartment.

Question 116.

Why not just fit the best available protection in each location?

Answer:

- The best possible protection, designed to prevent any ignition source from reaching any type of material would be too expensive and be over engineered.

- Classification societies and legislation take the approach that each enclosure should meet the requirements of the compartment it is installed in and no more.

Question 117.

How would you perform a maintenance test of a HV transformer?

Answer:

- I would obtain from the planned maintenance system, all relevant HV documents and history for this task.

- The authorising officer and I would both plan a switching scheme to make safe the area to be tested.

- Together we would implement the switching scheme to isolate the area under test.

- Warning signs would be fixed.

- The isolated area would be proven dead, the circuit main earth will be switched in.

- All switches used to isolate and earth this area to be locked out with suitable padlocks.

- The authorising officer would raise the Permit to Work which he and I would sign.

  The above switching scheme for a HV transformer would include the following steps:

- I would open, lock out and tag the breakers on either side of the transformer.

- I would, under direction of the authorising officer apply main circuit earths to the transformer.

- I would test that the transformer primary and secondary windings were dead using the approved HV indicator, proving the indicator is working before and after each test.

- With the transformer isolated, earthed and dead, I could perform the maintenance tasks.

- If the transformer used forced cooling, I would action the checks on the cooling system: temperature sensors, fans and pumps, including alarms and trips.

Question 118.

What tests would you perform on this transformer?

Answer:

- I would check the insulation resistance and polarisation index on each phase winding to earth.

- I would check the transformer cooling system.

- I would follow the tests mandated within the SMS using the approved equipment.

Question 119.

Why do we earth a HV system prior to test?

Answer:

- HV systems, due to their very high insulation resistance, high voltage and capacitance, retain a high voltage for a long time.

- Due to the very high value of the resistance, HV wiring can develop an induced EMF from stray HV fields from adjacent wiring.

- This high voltage can be lethal for days even when the device is fully isolated.

- This voltage is stored in the capacitance of the cabling since a capacitor is created when two conductors are separated by an insulator.

- So we have capacitance from phase 1 (L1) to L2, L1 to L3 and L2 to L3. Also from L1, L2 and L3 to earth.

- This is why we keep the main circuit earth(s) and additional earths if instructed, connected for as long as possible and only remove individual earths, as instructed for the duration of a test.

Question 120.

What is polarisation index and why do we use it?

Answer:

- Polarisation Index (PI) is used when testing HV resistance (typically of motors, transformers and cables) because it gives a better *view* of the effectiveness of the HV insulation.

- A special PI tester is used. When safely connected to the conductors under test, the tester measures the resistance of the subject under test after one minute and again after 10 minutes. Dividing the 10 minute resistance reading by the 1 minute resistance reading gives us the polarisation index.

- It should be greater than 2.0

Question 121.

Polarisation Index sounds complicated, is it necessary?

Answer:

- Polarisation Index is normally better than a single megger reading.

- A single megger reading will vary with temperature, the type of the insulation and the value to be read.

- At the start of the PI test, the tester applies the HV to the test connections.

- An initial surge of current flows, charging up the capacitance formed by the conductors and insulator.

- After one minute the initial charging current surge has reduced and the first resistance measurement is taken.

- With the voltage still applied, a delay of ten minutes is specified to allow for the polarisation of atoms within the insulation.

- The second measurement is then taken.

- The circuit under test requires earthing to discharge any energy stored during the application of the test voltage.

Question 122.

Take me through entry into an Enclosed Space?

Answer:

- I would consult the planned maintenance system for the necessary permits and history.

- I would consult senior members of the crew.

- With the permits completed and posted, the appropriate team assembled, Bridge and ECR informed.

- The space will be opened up, pipes closed and blanked off, locked and tagged, to prevent any possible fluid entry.

- The space will be forced ventilated for 24 hours prior to entry and internal lighting rigged.

- If possible, ventilation should be forced into one opening and exhausted via a second opening.

- After 24 hours, the compartments atmosphere is tested for three categories:

- Oxygen: 20.8%.

- Flammable gases and vapours: as close to zero as possible.

- Toxic gases: as low as is reasonably possible.

- As per the permit to work, the following will be available:

- An attendant situated by the entrance with tested communications to the work party and bridge team.

- Breathing apparatus to hand and tested.

- Lifeline.

- A rescue harness should be worn by each member of the work party.

- A personal gas detector should be worn.

- A rescue team should remain on standby for the duration of entry.

- After the work is complete, the work party exits with all equipment, compartment closed off, locks and tags removed, permits closed down and the SMS is updated.

Question 123.

How do you check a BA set is fit for purpose and ready for use?

Answer:

- I would check the external condition of the harness, pressure pipe and mask for damage and wear.

- Slowly open the cylinder valve all the way, and check that it contains 200 bar pressure.

- Close the cylinder valve and check that the pressure does not drop more than 10 bar within 1 minute.

- Slowly vent the pressure via the demand valve, checking that the low air alarm operates at 55 Bar.

- Don the mask, breath in and ensure the mask fits and has no leaks.

- Open the cylinder valve fully, breath normally, operate the bypass valve and check for free flowing air.

- Close bypass valve.

- Remove mask.

- Close cylinder valve.

Question 124.

Do you ever have rescue drills from an enclosed space?

Answer:

- Yes, the Code of Safe Working Practices informs us we need to have a rescue drill at least every 2 months.

- These rescue drills check we have the correct equipment, can use it correctly, that the entrance and exit method works.

- It gives us the opportunity to practice in a different enclosed space each time.

Question 125.

How could a sensor transmit its measurement to a controller or display?

Answer:

- A sensor could transmit its reading by:

- A 2 wire current system called *4 to 20mA loop* or

- A digital network such as Fieldbus, Profibus, Modbus or CAN bus.

- 4 to 20mA is simple, but digital networks allow two way communications so reliability data can be collected which can be used to reduce unscheduled maintenance.

Question 126.

Why are 4 to 20mA loop systems popular?

Answer:

- They are simple and reliable.

- Any value a sensor transmits, (temperature, pressure, RPM etc.) as a current, is guaranteed to arrive at its destination and a broken wire can be detected immediately.

- If you transmit a voltage, you can never guarantee the voltage received will be the same value as that transmitted due to volts drop along the wiring.

Question 127.

I noticed you did not mention the ethernet when you mentioned digital networks, why?

Answer:

- Ethernet technology is not a realtime system.

- Data transmitted via ethernet is not guaranteed to arrive at its destination in a timely manner.

- Field Bus, one of the more popular digital control networks, guarantees each packet of data will arrive at its destination within 10 milliseconds or so.

- Digital control networks are designed for situations where realtime control is demanded.

- A standard laptop connected to an ethernet control system in the engine room can easily overload and damage the control network.

- High Speed Ethernet (HSE) is being introduced for certain controller to controller communications but individual sensors and actuators still use Fieldbus.

Question 128.

How would you test a bilge high level alarm?

Answer:

- Depending upon its type I would:

- Lift the float the correct distance to trigger the alarm;

- Remove sensor and place in bucket of water.

Question 129.

What do we do and why, when we go on standby for leaving and entering port?

Answer:

- An additional generator is started and placed on the board.

- The second steering gear pump is started.

- The additional generator is used to reduce the likelihood of a complete electrical failure.

- The additional steering pump enables steering to be maintained during the failure of the either steering gear pump. It removes the possibility of an automatic changeover failure.

- In both cases, the additional hardware in use minimises any possibility of loss of functionality during complex vessel manoeuvres.

- A senior engineer (Chief or second) will be in attendance in case any unexpected incident occurs.

- The electrical officer and or ETO will be available depending upon the requirements of the company.

Question 130.

Tell me about four pillars of IMO?

Answer:

- IMO: International Maritime Organisation is to provide an international standard for the safe management and operation of ships and for pollution prevention.

- The four pillars of IMO are:

- SOLAS: Safety of Life at Sea,

- MARPOL: Prevention of Pollution from Ships,

- STCW: Standards of Training, Certification and Watchkeeping for Seafarers,

- MLC: Maritime Labour Convention.

Question 131.

Tell me about MARPOL?

Answer:

- MARPOL: *International Convention for the Prevention of Pollution from Ships* is the convention that aims to eliminate pollution by oil or other noxious substances at sea.

- It controls if and when substances can be emitted from the vessel in port or at sea, into the sea or, into the air.

- Over time, more stringent limits are being applied to discharges.

- It is now common for vessels to keep all effluent on board for safe disposal at the next port.

- Oily water can be discharged *en route*, if the discharge is filtered, no cargo oil allowed and oil content of discharge is less than 15 parts per million.

Question 132.

Can you explain the International Safety Management Code
(ISM)?

Answer:

- Its aim is to *ensure safety at sea, prevention of human injury
  or loss of life, and avoidance of damage to the environment, in
  particular to the marine environment, and to property.*

- The ISM code is legislation that compels shipping companies to
  install and operate Safety Management Systems (SMS) on board
  each vessel.

- The SMS onboard, is required provide a step by step guide for
  all operations on the vessel.

- Which include all planned maintenance tasks.

- Records are kept and inspected regularly by the company, Port
  State Control and others.

Question 133.

What is SOLAS?

Answer:

- SOLAS or *International Convention for the Safety of Life at Sea*
  is an international treaty that sets standards in the construction
  and operation of vessels.

- It covers a wide range of provisions in 14 chapters, including:

- Ship construction;

- Carriage of cargoes;

- Carriage of Dangerous goods;

- Requires the ISM to be implemented;

- Requires that GMDSS systems be installed, tested and maintained.

- Requires the ISPS code to be implemented.

Question 134.

What is ISPS?

Answer:

- ISPS or *The International Ship and Port Facility Security Code* establishes an international framework for cooperation between Governments, Government agencies, local administrations, shipping and port industries, in assessing and detecting potential security threats to ships or port facilities used for international trade.

Question 135.

What do you understand about rest hours detailed within the Maritime Labour Convention?

Answer:

- The maximum hours of work shall not exceed 14 hours in any 24 hour period and 72 hours in any 7 day period.

- The minimum hours of rest shall not be less than 10 hours in any 24 hour period and 77 hours in any 7 day period.

- If hours of rest are divided they shall not be divided into more than two periods, of which one, shall be a minimum of 6 hours.

- The interval between consecutive periods of rest shall not exceed 14 hours.

- The wording of the MLC code causes some ambiguity of its meaning.

Question 136.

Tell me about the Ballast Water Management Convention?

Answer:

- The BWMC demands that all vessels will eventually by fitted with ballast water treatment systems to kill any organisms in the ballast water.

- This is to prevent the uptake of ballast water containing live microorganisms in one port, with their discharge in a distant port, spreading these microorganisms that may have never existed in the new area before.

- Invasive species entering a new area can cause considerable damage to existing marine life.

- Each cubic metre of discharge shall contain less than 10 viable organisms equal to or greater than 50 micrometres in size

- *AND*

- Less than 10 viable organisms per millilitre between 10 micrometres and 50 micrometres in size.

Question 137.

What are the main methods of Ballast Water treatment?

Answer:

- Ultra Violet (UV), Chemical, Electrical and Filtering

- It is normal for multiple methods to be used to ensure compliance.

- Filtering with Ultra violet, filtering with Chemical and Filtering with electrical.

- Filtering removes large bodies from the water stream but many live organisms remain.

- Ultra violet uses a cylindrical bank of UV tubes radiating on a central clear pipe through which the ballast water flows.

- Chemical systems use a supply of a biocide, such as Chlorine or its compounds, hydrogen peroxide or ozone which are fed into the ballast water flow.

- Electrical systems discharge a large current pulse through the ballast water causing a rapid rise in temperature and pressure to kill the microorganisms.

- New systems are being approved constantly.

Question 138.

Your Generator is rotating but producing 0 volts, no electrical or electronic problems can be found, what would you do?

Answer:

- If no electrical problems can be found, I would flash the generator.

Question 139.

How would you Flash a brushless self excited Generator?

Answer:

- With the prime mover stopped, turning gear/air start valve/heaters locked out and tagged,

- Main breaker open, locked out and tagged,

- I would check that the AVR can withstand the flashing voltage, if not, I would disconnect the AVR from the exciter stator windings.

- I would check that the diode pack can withstand the flashing voltage, if not, I would disconnect or fit a shorting link across them.

- I would connect a 12V battery briefly across the exciter stator windings.

- This would re-magnetise the exciter stator.

- I would then reverse the safety procedure.

Question 140.

Can you tell me about problems with brushless self excited generators and workarounds?

Answer:

- A simple self excited generator takes power from the main output and feeds it via the AVR to the field windings of the exciter.

- If a short occurs to the main bus bar outputs somewhere on the ship, we expect the main circuit breaker to trip, open, protect the generator and loads.

- Since the output bus, which is shorted, feeds the exciter, the exciter is no longer able to supply sufficient current to the main generator rotor, causing the generator voltage output to collapse.

- This collapse can happen before the main circuit breaker has tripped.

- The reverse power trip then kicks in as the generator output is zero, but the breaker is still closed.

- Due to the AVR/exciter dependence upon the output voltage, these generators can be sensitive to harmonics caused by variable speed drives and other non-linear loads.

- These generators can increase the size of harmonics when driving variable speed drives due to the drives taking its power on short chunks, causing the excitation voltage to fluctuate in sympathy.

- Simple brushless self excited generators can be improved by replacing the exciter with a permanent magnet (PM) exciter.

- A PM exciter generates a voltage when ever the generator is rotating, enabling the generator output to remain long enough that the main circuit breaker operates correctly during short circuit conditions.

- Generators are required to deliver a minimum of three times the rated current at zero output voltage for two seconds (short circuit conditions) and 1.5 times the rated current for thirty seconds at full voltage (overload conditions).

Question 141.

Can you sketch the layout of a permanent magnet brushless generator please?

Answer:

- The brushless generator with PM exciter consists of the main alternator, an excitation alternator and a permanent magnet (PM) generator sharing a common shaft.

- The PM generator consists of a rotating permanent magnet mounted on the common generator shaft.

- Whenever the common shaft is rotated by the prime mover, voltage is produced by the PM generator.

- The output of the PM generator is fed to the AVR.

- The AVR, comparing the set point (SP) e.g. 440V with the output of the main generator and adjusts (or modulates) the voltage and therefore, the current through the exciter field windings.

- As the common shaft rotates, the rotor in the excitation alternator cuts the magnetic flux from the field windings, generating a three phase voltage.

- The voltage induced is rectified by the rotating diode pack.

- The DC from the diode pack is fed into the main alternator rotor creating a rotating magnetic field.

- This rotating magnetic field cuts the main alternator stator windings creating the 3 phase voltage output.

- The output from the stator supplies power to the bus bars and the AVR with a Measured Value (MV).

- The AVR is fed with a Set Point (SP) e.g. 440, 690 or 6600 etc.

- The AVR compares the MV to the SP and changes the current from the PM generator through the exciter stator windings to achieve the required voltage.

- The advantage of this type of generator is its improved short circuit handling capability and less prone to harmonic interference from variable speed drives.

- The sketch in figure 19 is suitable to answer this question.

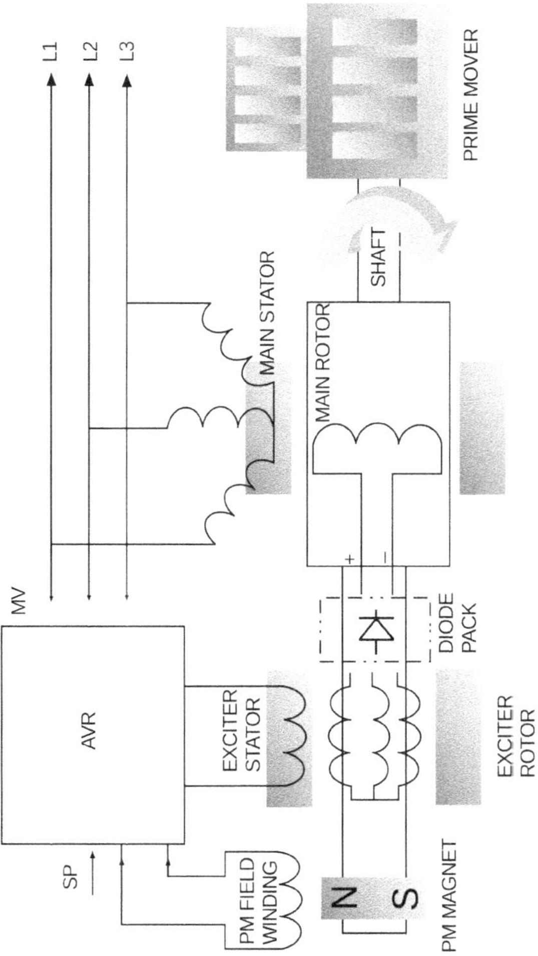

Figure 19: *Sample Brushless Alternator with PM Exciter.*

Question 142.

What can you tell me about Long-Range Identification and Tracking (LRIT)?

Answer:

- Long-Range Identification and Tracking (LRIT) is a system where the vessel automatically transmits its position to shore authorities on a regular basis.

- These days, it is common to have a self contained satellite based transmitter/receiver/satnav located in a sealed box mounted above the bridge, although any GMDSS MF/HF or satellite system could be used.

- The system transmits the vessels location every 6 hours. Shore stations can request more frequent position reports.

- It is used to give authorities enhanced situational awareness of shipping movements.

- It is unwise to switch this LRIT system off. You will bring your vessel to the immediate attention of international authorities!

Question 143.

Can you sketch and describe a UPS please?

Answer:

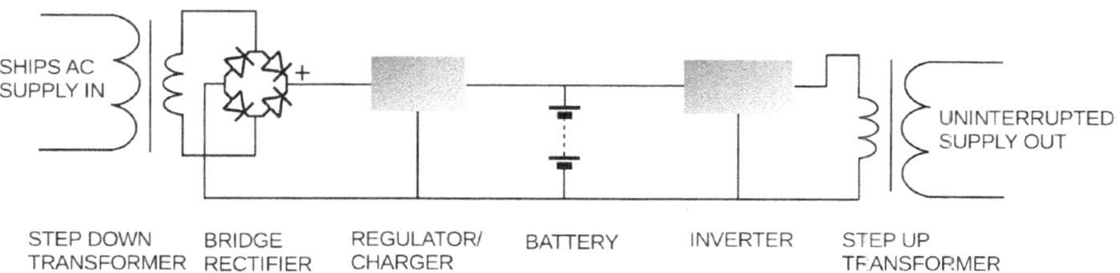

Figure 20: *UPS block diagram.*

- In figure 20 the left hand side shows a mains step down transformer, changing the 220VAC to 28VAC

- This is followed by a fullwave bridge rectifier giving us 28VDC.

- The 28VDC is smoothed and fed into the regulator system which ensures that the battery is fully charged at all times and feeds 24VDC to the inverter.

- The inverter transforms the DC voltage back to AC.

- The final block is the step up transformer to give the output voltage of 220VAC.

- The battery is constantly available to provide 24VDC power to the inverter.

- With a fully charged battery, if the 220VAC ships supply fails, the battery voltage ensures the inverter continues to generate AC.

- The battery can be internal to the UPS if the battery is small or can be mounted external to the UPS if the battery is large.

Question 144.

What how and why is steering gear tested?

Answer:

- SOLAS informs us we have to test our steering gear within 12 hours before departure.

- This includes both main and auxiliary steering gear.

- We check the rudder can travel from 35° to 30° within 28 seconds. In both directions.

- We check bridge to steering flat communications and rudder angle indicators.

- We check manual steering and emergency steering.

- We check for leaks.

Question 145.

What can you tell me about 1:1 ratio transformers?

Answer:

- 1:1 or one to one transformers can be used for:

- Isolation - small power transformers can isolate an electric hand tool user from the possibility of electric shock to earth.

- Isolation - transformers prevent earth faults from propagating along a distribution system.

- Harmonic reduction - the inductance property of a transformer attenuates or reduces high frequency harmonics passing through.

- Many variable speed drives are fed via a 1:1 transformer to reduce the harmonics from the motor controller being transmitted into the ships power distribution system.

Question 146.

What signals and information are fed to a radar display and why?

Answer:

- The gyro produced ships heading is fed to the radar display to enable a *North up* display. This gives the navigator, the same radar orientation as a paper chart.

- The radar display can be fed with data from the AIS, to give the navigator a *one screen view* of raw radar data and AIS symbols.

- Charts can be loaded into a modern radar to combine raw radar data and chart data.

- Also, the speed log, is fed to the radar display to enable *true motion* to be used.

- The heading and speed can be received from GPS as well.

- ARPA anti collision information is generated within the radar and can be selected for view.

- Most new builds use what are called multifunction displays (MFD).

- Which means the navigator can select any combination of navigational data for each display:

- Typically, ECDIS, radar and AIS on one display.

Question 147.

What points of concern and technical considerations with respect to lithium-ion batteries can you tell me about?

Answer:

- Lithium-ion batteries are the most power dense batteries currently available.

- MGN 550 (M+F) lists a number of concerns about Lithium-ion batteries on board a vessel. These include:

- Correct temperature monitoring and connection to the vessels fire alarm system.

- Monitor the charging and discharging process to prevent damage to the battery.

- A battery management system will achieve the above.

- Many modern, large installations are monitored in realtime from ashore by a battery specialist to increase the safety of the installation and achieve maximum battery life.

- Since each cell contains fuel and oxidiser it extremely difficult to extinguish. It is currently usual to use a high pressure water spray to cool adjacent cells to prevent those cells experiencing thermal runaway.

- Chemistry varies between battery manufacturers so each installation needs a dedicated fire fighting plan.

Question 148.

What is the difference between a HV permit to work and a Sanction to test?

Answer:

- A HV Permit to Work, documents the planning of isolation and making safe part of a HV system for maintenance.

- A HV Sanction to Test, documents and permits the removal of a safety earth for the purpose of conducting a measurement on an isolated and made safe HV system.

- During the operation of a HV permit to work, earths can not be removed.

Question 149.

How would you add a new printer to the computer network on board ship?

Answer:

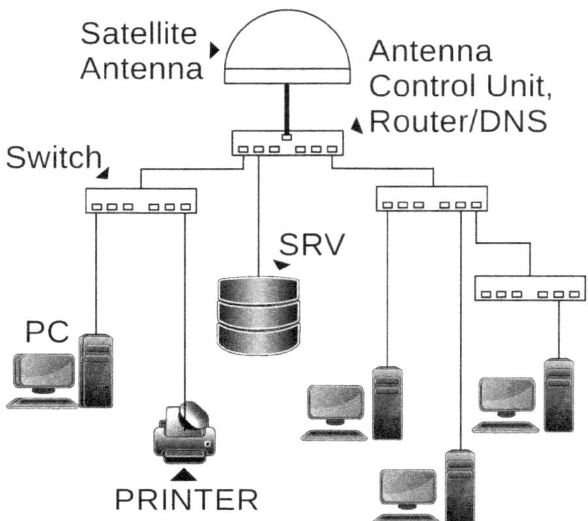

Figure 21: *New printer added to ships IT network.*

- I would confirm that the printer mains input was compatible with the ships domestic supply then connect it to the supply,

- I would connect the printer to the network using an RJ45 to RJ45 cable to an active wall socket or an adjacent network switch,

- Powering up the printer, I would wait a few minutes then run the *add printer* software on any PC that needed access to this new printer. Figure 21 shows a simplified network layout.

- If this type of printer had been installed on the network before, a driver should be available on the file server.

- If it was the first time this model of printer had been used on this network I would use the software driver disk that accompanied the printer.

118

- This would save the cost of downloading the driver software over the satellite connection.

Question 150.

What is GMDSS?

Answer:

- The Global Maritime Distress and Safety System (GMDSS) is a system that allows vessels to send and receive distress, urgency, safety and other essential messages.

- It specifies different radio equipment requirements dependent upon which area of the world the vessel trades.

Question 151.

Can you explain the area approach to GMDSS?

Answer:

- The system divides the oceans into four areas: A1, A2, A3 and A4. This allows the vessel equipment fit to suit the area the ship normally trades in.

- Area A1 is coastal, typically up to 20 to 30 miles from the coast which has a continuous watch via a DSC.

- Area A2 typically covers up to 150 miles from the coast using an Medium Frequency (MF) transmitter/receiver with DSC.

- Area A3 is between 70 degrees above and below the equator. It requires access to a Inmarsat geostationary satellite terminal with DSC.

- Area A4, which includes both poles, covers the remaining areas. This requires access to HF transmitter/receiver with DSC.

Question 152.

What is DSC?

Answer:

- Digital Selective Calling (DSC) is a function built into a radio system that allows the calling of ships and coast radio stations by entering an MMSI number.

- It allows the automatic transmission of a distress call which includes the vessels position.

- It maintains a DSC watch to enable the ships crew to respond to calls from other ships and coastal stations.

- All of your GMDSS equipment contains a DSC.

- The MMSI is a Maritime Mobile Service Identity number that uniquely identifies each vessel.

- It allows each ship to be individually called.

- Group MMSIs can be setup to allow multiple ships to be alerted.

Question 153.

What are the requirements for emergency power for the GMDSS installation?

Answer:

- If the vessel has an emergency generator, the GMDSS equipment shall be powered for a minimum of 1 hour by battery.

- If the vessel does not have an emergency generator the GMDSS equipment shall be powered for a minimum of 6 hours by battery.

Question 154.

Tell me about GMDSS battery maintenance?

Answer:

- GMDSS battery conditions requires daily, monthly and annual checks.

- Daily: Switch to battery power only. Record the voltage indicated on the GMDSS equipment. Then switch on a normal load such as holding down the transmit button on an empty channel and record the battery voltage. The voltage should drop by less than 10% of the offload voltage. This demonstrates the battery is charged, healthy and the wiring between battery and the equipment is not faulty.

- Monthly: If it is possible, the specific gravity of the battery electrolyte should be measured and recorded. Most modern batteries are sealed and this is not possible. The cleanliness of the batteries, terminals and wiring should be checked and recorded.

- Annually: The capacity of the batteries should be checked by discharging the batteries at the appropriate rate, to meet the 1 hour or 6 hour requirement to ensure the batteries will work as intended during a blackout. The battery should then be charged by the usual charging equipment and verified it reaches the required state of charge within 10 hours. This must be done in port when the GMDSS is not required. Care must be exercised to ensure a damaging deep discharge does not exceed the batteries capability and reduce its service life.
UPS used for GMDSS equipment should use its built-in battery discharge facility to monitor the condition of the battery. Only to be done in port.
Equipment manuals and previous log entries will guide you since new technology can appear at anytime.

Question 155.

Why do we use oil mist detectors in main engines?

Answer:

- Large Diesel engines use Oil Mist Detectors to raise an alarm or cause a shutdown if an oil mist is detected in the crankcase or other monitored area.

- Oil mist is created by hot spots within the engine. If the concentration of oil mist reaches the lower explosive level then an explosion can occur.

- Oil Mist Detectors are continuously fed with air sucked out of each crankcase space in turn.

- The air is fed into the detector which measures a reduction in light passing through a tube. A reference tube is used to generate a signal for comparison.

- If the level of oil mist goes above allowable limits an alarm and shutdown are triggered.

Question 156.

Can you sketch an Oil Mist Detector?

Answer:

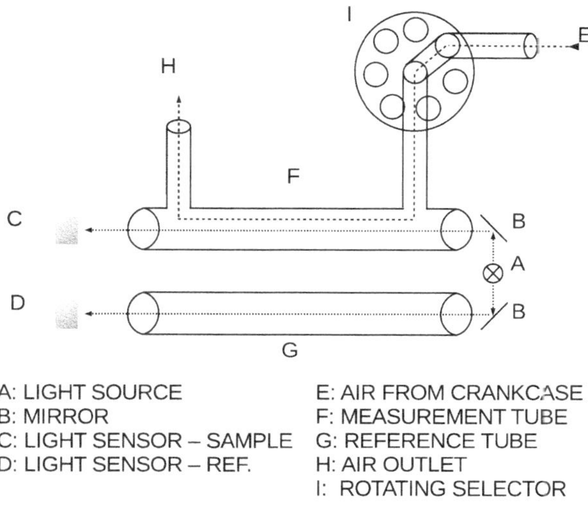

Figure 22: *Oil Mist Detector.*

- Figure 22 shows a traditional oil mist detector with one central sensor connected to multiple engine compartments with pipes which allow the sampled air to be sent to a single detector.

- Modern systems have an individual sensor in each compartment being sampled.

Question 157.

Can you sketch the droop curves of two paralleled generators?
What happens if you increase the fuel into one generator?

Answer:

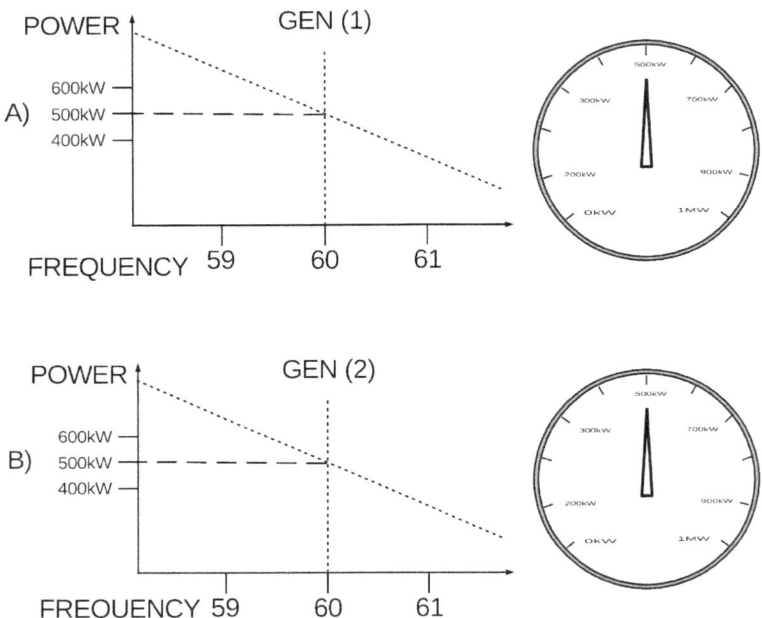

Figure 23: *Two generators paralleled, sharing a 1MW load.*

- Figure 23 shows Generator (1) and Generator (2) in parallel running at 60Hz sharing 50% of the load each.

- The sloping line is the droop characteristic or droop curve. It indicates that a generator will be forced by its governor to run slightly slower if the generator load is increased.

- As they are synchronised together, if one generator is forced to consume more fuel, the other will reduce its fuel use by its governor action, controlled by its own governor droop curve.

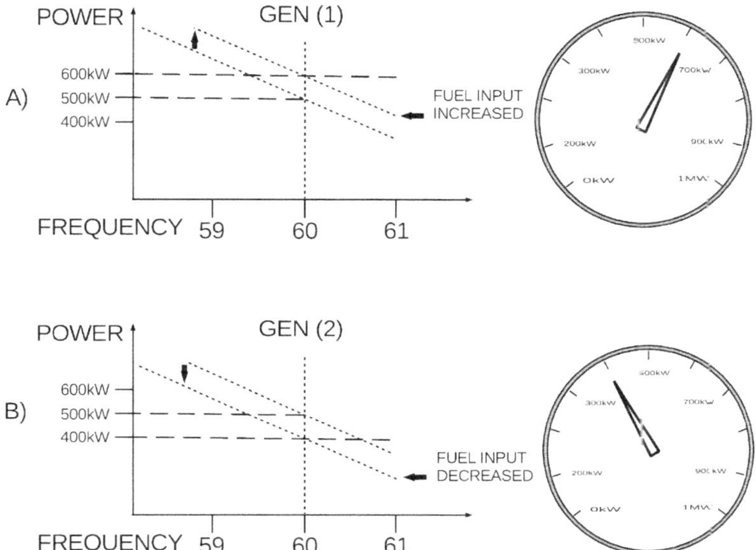

Figure 24: *Generator (1) setpoint increased causing more fuel to be used. Which causes Generator (2) to reduce fuel by its governor droop.*

- In figure 24 Generator (1) has been fed with more fuel by increasing the RPM or frequency setting.

- We can see the droop characteristic line has been increased in height indicating more fuel/power.

- Since the load has not changed, Generator (1) will try to speed up slightly. Generator (2) governor notes the slight increase in speed/frequency and, following the droop line, reduces the fuel into generator (2).

- The two black arrows on the right of the diagram indicate the movement of the droop curve. (1) rising due to operator action and (2) falling due to governor action.

- If the governor (2) was not following a droop curve, Generator (2) would have a frequency slightly higher that 60Hz. Perhaps 61Hz. By forcing generator (2) to use slightly less fuel, we would settle the frequency of both generators to 60Hz.

- Remember, the governor setpoint is what you want, say 60Hz. The governor adjusts the fuel rack to try to achieve 60Hz. If the load or the other generator governor changes, our governor will adjust the fuel rack again, following the droop curve. In this way, two or more generators can coexist, driving the bus bars, at the same frequency, with each generator governor adjusting its own fuel rack to compensate for any changes.

- Remember, both generators are synchronised together, so they always run at the same frequency.

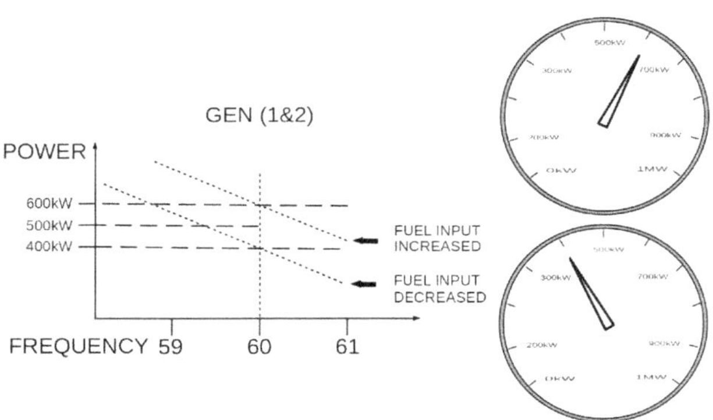

Figure 25: *Generator (1) setpoint increased causing more fuel to be used.*

- In figure 25 Generator (1) is shown producing 600kW at 60Hz and Generator (2) is shown producing 400kW at 60Hz.

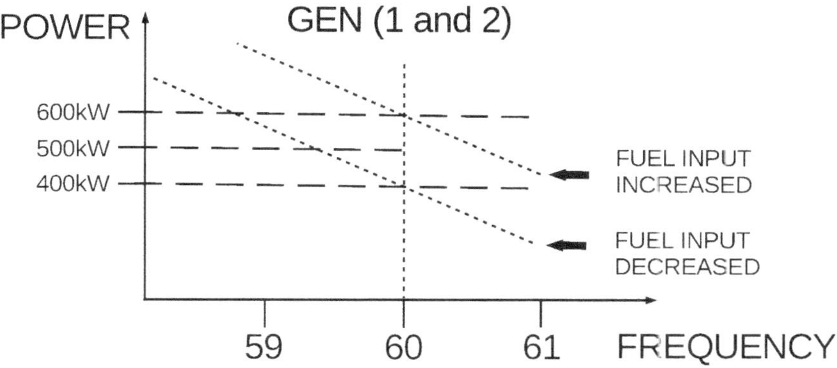

Figure 26: *Generator (1) setpoint increased causing more fuel to be used.*

- Figure 26 shows both droop lines on the same chart, without the dials, looking more representative for a sketched answer.

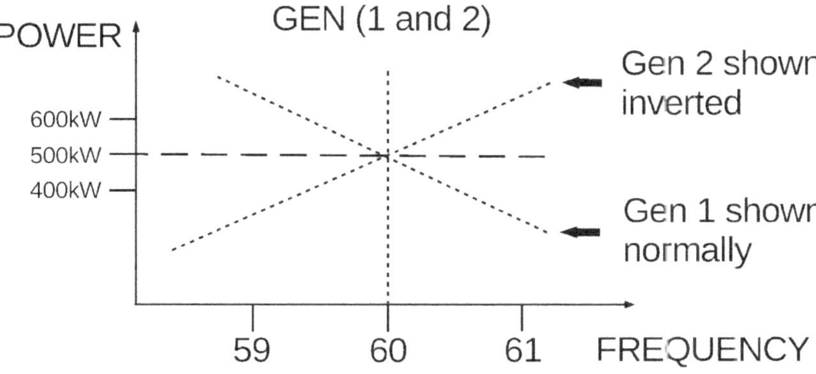

Figure 27: *Generator (2) droop curve inverted to show at equal power, the droop curves cross at 500kW.*

- In figure 27 Generator (1 and 2) can be seen with Gen (2) curve inverted or turned upside down. It shows the droop curves crossing each other at 60Hz. Some like to see an inverted droop curve giving a crossing of the lines but I do not think it is very helpful in increasing understanding.

- In figure 28 we can see that generator 2 droop curve has been inverted and the load share is 600kW and 400kW. Note, that the droop curves do not cross at the correct frequency of 60Hz which why I do not like the inverted curve display.

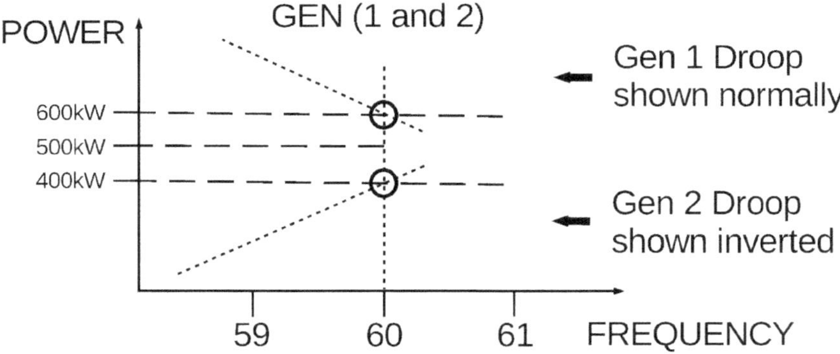

Figure 28: *Generator (1) setpoint increased causing it to supply more power at 60Hz.*

Question 158.

Can you explain how a hand cranked Megger works?

Answer:

- A Megger Tester is a make of high voltage insulation tester.

- The hand cranked version consists of a hand operated DC generator that typically produces 500V DC.

- It has a resistance display which indicates the resistance measured.

- To cope with the variable voltage generated by different hand cranking speeds, both the applied voltage and subsequent current are measured and the resulting resistance depends upon both the voltage and current.

- Using ohms law, R = V/I.

- To achieve this the display meter contains two moving coils mounted on the same shaft, with the pointer at the end of the shaft. One coil for the current to be measured and one coil for the voltage used to make that current flow. Both coils works in opposition to each other. By suitable design of the coils, number of turns and size of each coil an accurate reading can be obtained.

- The higher the current measured indicates a lower resistance so the meter pointer moves further.

- The higher the generated voltage would cause the resistance to read incorrectly lower, so the voltage coil torque interacts and partially cancels the current coil torque.

- By careful balance of forces an accurate reading can be obtained.

- A coiled return spring is fitted to the shaft to ensure an infinity reading is shown with no connection to the leads.

- Figure 29 shows the hand cranked generator providing 500VDC across the voltage coil (V) and feeds current through the current coil (I) through the test subject.

- Visualise the circuit with the voltage coil missing. It is then just an ordinary resistance meter. But with poor accuracy due to the poorly regulated DC voltage coming from the hand cranked generator. With the crank turning, approximately 500V DC will be applied to the test subject. Current will flow through the test subject and through the current coil (I). This will cause the coil to develop some torque and move the meter pointer. If the 500V DC was a bit low today, the pointer will not move a much giving inaccurate readings and vice versa. Adding a voltage coil (V) with corrects for varying voltages from the hand cranked generator ensures and accurate reading by producing the opposite torque. Higher voltage equals more current through the test subject, more current through the current coil (I) BUT more opposite torque from the voltage coil.

- It is difficult to find hand cranked test equipment now. New equipment is mostly fully electronic.

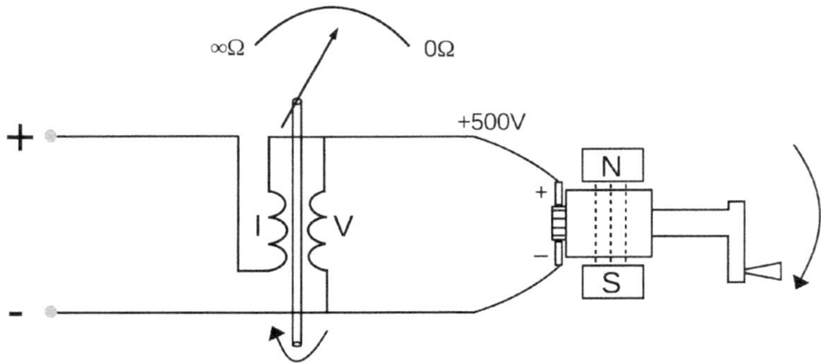

Figure 29: *Internals of a hand cranked Megger insulation tester.*

Question 159.

What HV test equipment have you used?

Answer:

- Earthing spider, Proving unit, Discharge Rod and HV Measuring indicator.

Question 160.

Can you tell me how you use a HV Measuring Indicator?

Answer:

- The HV measuring indicator was a seaward KDIE device. It came with a proving unit.

- I would prove the indicator was working correctly by connecting it safely to the proving unit, prove the functionality of the indicator.

- I then connected it to the live terminal to be proven dead and check the indicator.

- I would then re-prove the indicator by retesting with the proving unit.

- If the point tested was proven to be zero volts it would be safe to apply an earth.

- If the point was isolated but still had a voltage reading, I would use the discharge rod, to discharge any residual high voltage down to earth.

- I would then prove, test and prove then earth.

- The above would be done using the SMS job information to ensure maximum safety.

Question 161.

Can you explain the differences between the three main type of Smoke/Fire detectors?

Answer:

- Smoke detectors are mainly of three types: Ionising, optical and heat.

- Ionising consists of a small radioactive source mounted between two plates. The radiation ionises the air between the two plates and a small voltage across the plates causes a small current to flow. Smoke particles appearing between the two plates causes some of the ionised particles to be cancelled and reduces the flow of current which can be measured and tested for.

- Optical detectors consist of a light source and a photo cell. The source shines on the cell causing a small current to flow. If smoke partially interrupts the beam of light will trigger less current to flow causing an alarm.

- Flame detectors detect IR and/or UV light and monitors for flashing or changing light levels.

- Heat detectors consist of a temperature sensor such as a thermistor or PT100 which responds to temperature change.

- Some alarm sensor heads contain multiple sensor technology to increase the reliability of the detector and reduce the incidence of false alarms.

- The correct choice of detector is required to give rapid detection with the least number of false alarms.

3 Mock Interview 1

I was asked:

- Draw distribution diagram (examiner looked through my TRB whilst I did this).

- What protection does an electric motor have?

- Starters:

- What methods of starting are there? Why are there different methods?

- Starting currents - how large are they at starting for each method?

- Draw a Star-Delta starter? Explain the operation. What if all contactors remained closed?

- If we are getting lower current in Star why not continue running in Star? What will happen?

- Thermal overload?

- Describe Earth faults:

- Significance of earth faults?

- Why different from shoreside?

- Why do we need to find earth faults?

- What tells us there is an earth fault?

- Is the identification of phase significant?

- How do we find them?

- Explain *selective* isolation of circuits? - then led to talk about essential systems, redundancy and why run two steering gear motors together - speed of response?

- What causes intermittent earth faults? Where might motor get wet and dirty? What about on 230V system?

- How does a generator work?

- Generator types?

- What generator protection is there?

- Emergency generator regulations?

- How does the emergency generator get tested on-load and how does it start?

- Batteries and battery maintenance. UPS also.

- GMDSS and testing.

- Power Management Systems (PMS):

- One generator running in port and about to sail - PMS won't let you put another generator on the board because it doesn't need it - what do you do?

- PMS tries to put a generator on the board but keeps failing to synchronise - manual operation fails too - what could be the problem? Why no voltage?

- What protection on circuit breakers? Purpose of circuit breakers?

- Using governor controls to balance the load between generators but the load is not being transferred from one to the other - what is the problem?

- How to IR test a motor and how would you know it was good, min ohm value

- Difference between HV IR values and LV values

- Harmonics:

- What are they? What causes them? What is a likely source?

- Explained PWM from my vessel.

- What would you get on a large passenger ship?

- Explained Synchronous motors and their associated convertors

- Discussed the benefits of HV

- Hazardous areas:

- What are they?

- What equipment would you put in cargo pump room?

- What is a transformer

- What faults could lead to a motor drawing excess current

- Asked me about Frequency drives briefly.

- What type of equipment would use frequency drives

- Why use a frequency drive

- What other types of speed control are there?

4 Mock Interview 2

I was asked:

- Draw a ships electrical distribution system.

- Fire detection system and types of detectors.

- Procedure for CO2 release.

- Enclosed entry procedure and what gases you are looking to detect.

- What main engines were on-board?

- Steering gear types and steering gear alarms.

- Why HV is used on board?

- HV full isolation procedure.

- HV and LV breaker differences and operation and how to maintain them.

- Draw either Star-Delta or Auto-transformer and explain its operation and why it's used.

- How would you know the Star-Delta starter had changed from Star to Delta?

- What would I do if I was carrying out maintenance on a flameproof fitting?

- Radar maintenance procedure start to finish (working aloft, rescue plan etc).

- Describe how a variety of bridge equipment was connected to each other and what each device did.

- How does a generator work?

- Generator types.

- What generator protection is there?

- How to test reverse power?

- How does reverse power work?

- How to parallel two generators?

- How does differential protection for generator windings work?

- Emergency generator regulations.

- How does the emergency generator get tested on-load and how does it start?

- Explain different types of emergency starting methods.

- Generator circuit breaker won't close even though voltage, frequency and phase are all fine what are the likely problems.

- What batteries were on board and how were they tested, how often, what did they feed.

- Supplementary lighting regulations (these are the battery exit lights).

- What is a transformer and its function why VA ratings and not Watts?

- How to IR test a motor and how would you know it was good, minimum ohm value?

- Polarisation Index - what it is and what it means?

- Describe the difference between salient pole rotors and cylindrical rotors.

5 ETO Oral syllabus - MIN 654 (M) Issue Feb 2021

The latest version of the syllabus is available from the MCA website, search for *MIN 654 – Amendment 1 (M) MCA gov.uk*. It is updated from time to time.

MIN 654 (M) Feb 2021 is competency table is formated in four columns titled: Competence; Knowledge, understanding and efficiency; Criteria for evaluating competence; Further guidance for evaluating competence.

Due to the size of this book, I reformated the table into a series of four sections e.g (1/1, 1/2, 1/3, 1/4) up to (18/1, 18/2, 18/3, 18/4) representing each row of competency in the table. This book will concentrate on the core areas of knowledge required by someone who seeks an ETO CoC. This is what the examiner will concentrate on, to determine if you are safe and competent. See the following section that describes the minimum standards required.

Electrotechnical Officers (ETO) Certificate of competency Oral Exam Syllabus

Table A-III/6

Specification of minimum standard of competence for electro-technical officers

1. Every candidate for certification as Electro-Technical Officer shall be required to demonstrate the ability to undertake the tasks, duties and responsibilities listed in the 'competence column A-III/6 table.

2. The minimum knowledge, understanding and proficiency required for certification is listed in the 'Knowledge, understanding and proficiency' column of table A-III/6.

3. Every candidate for certification shall be required to provide evidence of having achieved the required standard of competency tabulated in the 'Criteria for evaluating competence' and 'Further guidance for evaluating competence' columns of table A-III/6.

Function: Electrical, electronic and control engineering at the operational level

1/1 Competence

Monitor the operation of electrical, electronic and control systems.

1/2 Knowledge, understanding and proficiency.

Basic understanding of the operation of mechanical engineering systems, including:

1. prime movers, including main propulsion plant;

2. engine-room auxiliary machinery;

3. Steering systems;

4. Cargo handling systems;

5. Deck machinery;

6. Hotel systems.

Basic knowledge of heat transmission, mechanics and hydromechanics.

Knowledge of:

Electro-technology and electrical machines theory.

Fundamentals of electronics and power electronics.

Electrical power distribution boards and electrical equipment.

Fundamentals of automation, automatic control systems and technology.

Instrumentation, alarm and monitoring systems.

Electrical drives.

Technology of electrical materials.

Electro-hydraulic and electro-pneumatic control systems.

Appreciation of the hazards and precautions required for the operation of power systems above 1,000 volts.

1/3 Criteria for evaluating competence

Operation of equipment and system is in accordance with operating manuals.

Performance levels are in accordance with technical specifications.

1/4 Further guidance for evaluating competence

Relevant aspects of candidates knowledge and experience in these areas is assessed in the oral examination.

Candidates knowledge, understanding and proficiency of electrical and electronic devices, electrical motors and generators, control systems and transducers, power electronics, pneumatics and hydraulics and transformers are evaluated.

Understanding of switchgear and protection of high voltage systems.

Excitation methods.

Thyristor controls etc.

Candidates knowledge of understanding and proficiency regarding: Switchboard safety devices and breaker types;

Understanding of earth faults;

MEGGER readings;

Dangers of electric arc flash.

2/1 Competence

Monitor the operation of automatic control systems of propulsion and auxiliary machinery.

2/2 Knowledge, understanding and proficiency

Preparation of control systems of propulsion and auxiliary machinery for operation.

2/3 Criteria for evaluating competence

Surveillance of main propulsion plant and auxiliary systems is sufficient to maintain safe operation condition.

2/4 Further guidance for evaluating competence

Relevant aspects of candidates knowledge and experience in these areas is assessed in the oral examination.

3/1 Competence

Operate generators and distribution systems

3/2 Knowledge, understanding and proficiency

Coupling, load sharing and changing over generators.

Coupling and breaking connection between switchboards and distribution panels.

3/3 Criteria for evaluating competence

Operations are planned and carried out in accordance with operating manuals, established rules and procedures to ensure safety of operations.

Electrical distribution systems can be understood and explained with drawings/instructions.

3/4 Further guidance for evaluating competence

Understanding of paralleling and load sharing of generators.

Earth lamps.

4/1 Competence

Operate and maintain power systems in excess of 1,000 volts.

4/2 Knowledge, understanding and proficiency

Theoretical knowledge

High-voltage technology.

Safety precautions and procedures.

Electrical propulsion of the ships, electrical motors and control systems.

Practical knowledge

Safe operation and maintenance of high-voltage systems, including knowledge of the special technical type of high-voltage systems and the danger resulting from operational voltage of more than 1,000 volts.

4/3 Criteria for evaluating competence

Operations are planned and carried out in accordance with operating manuals, established rules and procedures to ensure safety of operations.

4/4 Further guidance for evaluating competence

Relevant aspects of candidates knowledge and experience in these areas is assessed in the oral examination.

Understanding of Switchgear and protection of high voltage systems.

5/1 Competence

Operate computers and computer networks on ships

5/2 Knowledge, understanding and proficiency

Understanding of:

1. main features of data processing;

2. construction and use of computer networks on ships;

3. bridge-based, engine-room based and commercial computer use.

5/3 Criteria for evaluating competence

Computer networks and computers are correctly checked and handled.

5/4 Further guidance for evaluating competence

Basic understanding of Local area Networking.

Lighting and control systems.

6/1 Competence

Use English in written and oral form.

6/2 Knowledge, understanding and proficiency

Adequate knowledge of the English language to enable the officer to use engineering publications and to perform the officer's duties.

6/3 Criteria for evaluating competence

English language publications relevant to the officer's duties are correctly interpreted.

Communications are clear and understood.

6/4 Further guidance for evaluating competence

Basic entry level requirements.

Oral exam.

7/1 Competence

Use internal communicationsystems.

7/2 Knowledge, understanding and proficiency

Operation of all internal communication systems on board.

7/3 Criteria for evaluating competence

Transmission and reception of messages are consistently successful.

Communication records are complete, accurate and comply with statutory requirements.

7/4 Further guidance for evaluating competence

Basic understanding of different marine vessel communication systems including emergency networks ones.

Communication battery and back up checks, records.

Function: Maintenance and repair at the operational level

8/1 Competence

Maintenance and repair of electrical and electronic equipment.

8/2 Knowledge, understanding and proficiency

Safety requirements for working on shipboard electrical systems, including the safe isolation of electrical equipment required before personnel are permitted to work on such equipment.

Maintenance and repair of electrical system equipment, switchboards, electric motors, generators and DC electrical systems and equipment.

Detection of electric malfunction, location of faults and measures to prevent damage.

Construction and operation of electrical testing and measuring equipment.

Function and performance tests of the following equipment and their configuration:

1. monitoring systems;

2. automatic control devices;

3. protective devices.

The interpretation of electrical and electronic diagrams.

8/3 Criteria for evaluating competence

Safety measures for working are appropriate.

Selection and use of hand tools, measuring instruments, and testing equipment are appropriate and interpretation of results is accurate. Dismantling, inspecting, repairing and reassembling equipment are in accordance with manuals and good practice.

Reassembling and performance testing is in accordance with manuals and good practice.

8/4 Further guidance for evaluating competence

Relevant aspects of candidates knowledge and experience in these areas is assessed in the oral examination.

Maintenance of electrical equipment and systems.

Control systems, transducers.

Understanding of electrical safety.

Candidates Knowledge, understanding and proficiency of electrical safety.

Understanding of Switchgear and protection of high voltage systems.

Understanding of basic fault finding, electrical motors and alternators.

9/1 Competence

Maintenance and repair of automation and control systems of main propulsion and auxiliary machinery.

9/2 Knowledge, understanding and proficiency

Appropriate electrical and mechanical knowledge and skills.

Safety and emergency procedures.

Safe isolation of equipment and associated systems required before personnel are permitted to work on such plant or equipment.

Practical knowledge for the testing, maintenance, fault finding and repair.

Test, detect faults and maintain and restore electrical and electronic control equipment to operating condition.

9/3 Criteria for evaluating competence

The effect of malfunctions on associated plant and systems is accurately identified, ship's technical drawings are correctly interpreted, measuring and calibrating instruments are correctly used and actions taken are justified.

Isolation, dismantling and reassembly of plant and equipment are in accordance with manufacturer's safety guidelines and shipboard instructions and legislative and safety specifications. Action taken leads to the restoration of automation and control systems by the method most suitable and appropriate to the prevailing circumstances and conditions.

9/4 Further guidance for evaluating competence

Relevant aspects of candidates knowledge and experience in these areas is assessed in the oral examination.

10/1 Competence

Maintenance and repair of bridge navigation equipment and ship communication systems.

10/2 Knowledge, understanding and proficiency

Knowledge of the principles and maintenance procedures of navigation equipment, internal and external communication systems.

Theoretical knowledge:

Electrical and electronic systems operating in flammable areas.

Practical knowledge:

Carrying out safe maintenance and repair procedures.

Detection of machinery malfunction, location of faults and action to prevent damage.

10/3 Criteria for evaluating competence

The effect of malfunctions on associated plant and systems is accurately identified, ship's technical drawings are correctly interpreted, measuring and calibrating instruments are correctly used and actions taken are justified.

Isolation, dismantling and re-assembly of plant and equipment are in accordance with manufacturer's safety guidelines and shipboard instructions, legislative and safety specifications. Action taken leads to the restoration of bridge navigation equipment and ship communication systems by the method most suitable and appropriate to the prevailing circumstances and conditions.

10/4 Further guidance for evaluating competence

Electrical systems in potentially explosive and gas hazardous areas.

11/1 Competence

Maintenance and repair of electrical, electronic and control systems of deck machinery and cargo-handling equipment.

11/2 Knowledge, understanding and proficiency

Appropriate electrical and mechanical knowledge and skills.

Safety and emergency procedures.

Safe isolation of equipment and associated systems required before personnel are permitted to work on such plant or equipment.

Practical knowledge for the testing, maintenance, fault finding and repair.

Test, detect faults and maintain and restore electrical and electronic control equipment to operating condition.

11/3 Criteria for evaluating competence

The effect of malfunctions on associated plant and systems is accurately identified, ship's technical drawings are correctly interpreted, measuring and calibrating instruments are correctly used and actions taken are justified.

Isolation, dismantling and re-assembly of plant and equipment are in accordance with manufacturer's safety guidelines and shipboard instructions, legislative and safety specifications. Action taken leads to the restoration of deck machinery and cargo-handling equipment by the method most suitable and appropriate to the prevailing circumstances and conditions.

11/4 Further guidance for evaluating competence

Relevant aspects of candidates knowledge and experience in these areas is assessed in the oral examination.

Fault finding

Electrical systems in potentially explosive and gas hazardous environment.

12/1 Competence

Maintenance and repair control and safety systems of hotel equipment.

12/2 Knowledge, understanding and proficiency

Theoretical knowledge:

Electrical and electronic systems operating in flammable areas.

Practical knowledge:

Carrying out safe maintenance and repair procedures.

Detection of machinery malfunction, location of faults and action to prevent damage.

12/3 Criteria for evaluating competence

The effect of malfunctions on associated plant and systems is accurately identified, ship's technical drawings are correctly interpreted, measuring and calibrating instruments are correctly used and actions taken are justified.

Isolation, dismantling and re-assembly of plant and equipment are in accordance with manufacturer's safety guidelines and shipboard instructions, legislative and safety specifications. Action taken leads to the restoration of control and safety systems of hotel equipment by the method most suitable and appropriate to the prevailing circumstances and conditions.

12/4 Further guidance for evaluating competence

Maintaining electrical equipment/systems, electrical motors and generators, electrical safety.

Function: Controlling the operation of the ship and care for persons on board at operational level

13/1 Competence

Ensure compliance with pollution-prevention requirements.

13/2 Knowledge, understanding and proficiency

Prevention of pollution of the marine environment.

Knowledge of the precautions to be taken to prevent pollution of the marine environment.

Anti-pollution procedures and all associated equipment.

Importance of proactive measures to protect the marine environment.

13/3 Criteria for evaluating competence

Procedures for monitoring shipboard operations and ensuring compliance with pollution-prevention requirements are fully observed.

Actions to ensure that a positive environmental reputation is maintained.

13/4 Further guidance for evaluating competence

Relevant aspects of candidates knowledge and experience in these areas is assessed in the oral examination.

IMO legislation including MARPOL and BWM Conventions.

14/1 Competence

Prevent, control and fight fire on board.

14/2 Knowledge, understanding and proficiency

Fire prevention and fire-fighting appliances.

Ability to organize fire drills.

Knowledge of classes and chemistry of fire.

Knowledge of fire-fighting systems.

Action to be taken in the event of fire, including fires involving oil systems.

14/3 Criteria for evaluating competence

The type and scale of the problem is promptly identified and initial actions conform with the emergency procedure and contingency plans for the ship.

Evacuation, emergency shutdown and isolation procedures are appropriate to the nature of the emergency and are implemented promptly.

The order of priority, and the levels and time-scales of making reports and informing personnel on board, are relevant to the nature of the emergency and reflect the urgency of the problem.

14/4 Further guidance for evaluating competence

Relevant aspects of candidates knowledge and experience in these areas is assessed in the oral examination.

15/1 Competence

Operate life-saving appliances.

15/2 Knowledge, understanding and proficiency

Life-saving

Ability to organize abandon ship drills and knowledge of the operation of survival craft and rescue boats, their launching appliances and arrangements, and their equipment, including radio life-saving appliances, satellite EPIRBs, SARTs, immersion suits and thermal protective aids.

15/3 Criteria for evaluating competence

Actions in responding to abandon ship and survival situations are appropriate to the prevailing circumstances and conditions and comply with accepted safety practices and standards.

15/4 Further guidance for evaluating competence

Relevant aspects of candidates knowledge and experience in these areas is assessed in the oral examination.

16/1 Competence

Apply medical first aid on board ship.

16/2 Knowledge, understanding and proficiency

Medical aid

Practical application of medical guides and advice by radio, including the ability to take effective action based on such knowledge in the case of accidents or illnesses that are likely to occur on board ship.

16/3 Criteria for evaluating competence

Identification of probable cause, nature and extent of injuries or conditions is prompt and treatment minimizes immediate threat to life.

16/4 Further guidance for evaluating competence

Relevant aspects of candidates knowledge and experience in these areas is assessed in the oral examination.

17/1 Competence

Application of leadership and teamworking skills.

17/2 Knowledge, understanding and proficiency

Working knowledge of shipboard personnel management and training. Ability to apply task and workload management, including:

1. Planning and co-ordination.

2. Personnel assignment.

3. Time and resource constraints.

4. Prioritization.

Knowledge and ability to apply effective resource management:

1. Allocation, assignment, and prioritization of resources.

2. Effective communication on board and ashore.

3. Decisions reflect consideration of team experiences.

4. Assertiveness and leadership, including motivation.

5. Obtaining and maintaining situational awareness.

Knowledge and ability to apply decision-making techniques:

1. Situation and risk assessment.

2. Identify and consider generated options.

3. Selecting course of action.

4. Evaluation of outcome effectiveness.

17/3 Criteria for evaluating competence

The crew are allocated duties and informed of expected standards of work and behaviour in a manner appropriate to the individuals concerned.

Training objectives and activities are based on assessment of current competence and capabilities and operational requirements.

Operations are planned and resources are allocated as needed in correct priority to perform necessary tasks.

Communication is clearly and unambiguously given and received.

Effective leadership behaviours are demonstrated.

Necessary team member(s) share accurate understanding of current and predicted vessel state and operational status and external environment. Decisions are most effective for the situation.

17/4 Further guidance for evaluating competence

Relevant aspects of candidates knowledge and experience in these areas is assessed in the oral examination.

18/1 Competence

Contribute to the safety of personnel and ship.

18/2 Knowledge, understanding and proficiency

Knowledge of personal survival techniques.

Knowledge of fire prevention and ability to fight and extinguish fires.

Knowledge of elementary first aid.

Knowledge of personal safety and social responsibilities.

18/3 Criteria for evaluating competence

Appropriate safety and protective equipment is correctly used.

Procedures and safe working practices designed to safeguard personnel and the ship are observed at all times.

Procedures designed to safeguard the environment are observed at all times.

Initial and follow-up actions on becoming aware of an emergency conform with established emergency response procedures.

18/4 Further guidance for evaluating competence

Relevant aspects of candidates knowledge and experience in these areas is assessed in the oral examination.

The above ETO Oral syllabus from MIN 654 Amendment 1 (M) issue February 2021 (C)Crown Copyright 2021.

6 The examiner

It may be helpful to understand the background of your examiner. He or she will be an expert. Depending upon which country you take the CoC interview, your examiner probably will have been (in order): a Chief Engineer, a superintendent and a surveyor. Remember, you are a trainee, they are the expert and your examiner. Do not try to bluff or blag them, they will see through you immediately. It is not a competitive interview, the examiner wants you to pass if you demonstrate competence and a good knowledge of safety, especially with the safety critical questions.

7 Your approach

You are taking a job interview which can set you up for life! Take it seriously. Dress smartly, sit up straight, be confident and give firm answers to each question. If you do not say a word or phrase, for example, *prove, test, prove*, the examiner can only assume that you do not know it and will rightly fail you. Do not think answers to yourself, say the answers out loud.

Most oral interviews take place online. You will have your PC or laptop setup and tested in advance. The room will be prepared so that when the examiner asks you to turn your computer around by 360 degrees, they will see a mainly plain room, with little if anything around that could conceal notes and answers etc. Doors and windows shut so no one can gain entry during your oral interview. Consider closing curtains or ensuring the camera on your PC is not facing the window. This could cause you to appear in silhouette which the examiner will not like! Currently the examiners use Microsoft Teams video conference software. This runs on most computers and can be run in a browser window at the time of writing, as well as a paid for application. Do check your system to make sure your video software runs smoothly.

On the big day, make sure you are well rested and alert. Dress in smart business attire. You will present any of your remaining

documents to the examiner when asked. They will scan them quickly
and start asking you questions immediately. Calmly answer each
question as completely as possible but do not go on. Just answer the
question. Depending on your answer and the confidence with which
you spoke will cause the examiner to either move onto the next topic
or dig a little deeper into your answer. Confidence is important. If you
make a mistake whilst answering a question say so immediately. There
is latitude on a technical answer BUT if you make a mistake during a
safety critical answer, your fate can be sealed immediately!

Safety critical answers are most important. Be very aware that the
examiner is trying to find out if you are a safe worker, safe to your
colleagues, safe to the ship and safe to the environment. If you do not
tell the examiner a word or phrase in answer to a question, they will
assume you do not know and you will be failed.

When reading this book, keep repeating each question, read it out
loud, write the question down, say the answer, write the answer down.
Repeat, repeat repeat. If you learn hard, you will pass more easily.
(That's the theory!)

What the MCA say in MIN 690 November 2024:

*The MCA uses Microsoft Teams for online face-to-face oral
examinations. Candidates will require access to Microsoft Teams
prior to the start of the examination. Candidates must log onto
the meeting at least 10 minutes prior to their designated start time.
If a candidate does not appear in the virtual waiting room within
five minutes after the scheduled examination start time, they will be
deemed to have failed. Candidates should take the examination in a
suitable room and have the necessary amenities. E.g., a clear desk,
comfortable chair, water, low ambient noise, and adequate lighting
with no glare. The device in use must have a camera and access to
a reliable, high-speed, internet connection (recommended minimum of
10mbps), enough for Microsoft Teams to be successfully supported.
Candidates are not permitted to use background effects, turn off
the microphone or wear headphones during any part of the online
oral examination. If, during the exam, there are connectivity or IT
equipment issues resulting in repeated disruption of the examination,
this will result in postponement of the examination and the candidate*

8 My approach

I will tell you about most topics but will lightly touch on the last part of the syllabus *(Controlling the operation of the ship and care for persons on board at the operational level)*. Apart from your direct knowledge of MARPOL, you have been on a series of short courses that cover these aspects. The examiner may ask you about the contents of the last section, but they know you have passed all of the industry standard short courses and you should be able to answer correctly. This is no excuse for you not to study the content of your short courses. You really have to have a full understanding of fire fighting, breathing apparatus, first aid and pollution control and prevention. Just keep in mind all of your short courses!

I reference official documents as necessary so you know where the *official answer* come from. Do check that we are both (you and me) are using the latest version of each referenced document. They are updated from time to time, the latest version becoming the *new answer*!

9 What next

Read the syllabus, noting any areas you feel weak in. Read the sample questions and answers. Any question that you did not immediately know the correct answer, mark the page with a cross. After you have been though all sample question and answers, you will now have a series of pages/questions with crosses marked on them. For each marked question/answer write out the question, read it out aloud, ask a colleague to read it out loud to you, then copy out the answer. Then read out the answer. Remember we are not learning parrot fashion or by rote, you need to understand each question and each answer. If you show a lack of confidence when speaking an answer, the examiner may explore this area of your knowledge in greater detail to make sure you really know it! Repeat, repeat and repeat!

Then move onto the sample mock CoC interviews. Try each one, if you experience any difficulty, repeat, repeat and repeat.

By the time you have read and understood the book a few times, your appointment for your CoC should be near.

Good luck

If you find any errors you wish to tell me about or have suggestions for new questions, please email me at:

steverichardseto@gmail.com

Index